How to Get More Miles per Gallon

How to Get More Miles per Gallon

An indispensable glove-compartment guide for every car owner in America with 282 tips to save you gas — and money

by Robert Sikorsky

ST. MARTIN'S PRESS

Copyright © 1978 by Robert Sikorsky
All rights reserved. For information, write:
St. Martin's Press
175 Fifth Avenue
New York, N.Y. 10010
Manufactured in the United States of America
Library of Congress Catalog Card Number: 78-3973

Library of Congress Cataloging in Publication Data
Sikorsky, Robert.
    How to get more miles per gallon.

Includes index.
    1. Automobiles—Fuel consumption.   I. Title.
TL151.6.S54   629.2'53   78-3973
ISBN 0-312-39589-2 cloth
ISBN 0-312-39590-6 pbk.
ISBN 0-312-39591-4 prepack

For my son Kyle, and all the children of his generation, with the hope that we have the foresight to conserve fuel today so they may know the pleasure and responsibility of driving a car in the future.

"There is no energy policy that will do as much as voluntary conservation."

—President Carter

"Nature never gives anything away. Everything is sold at a price."

—Ralph Waldo Emerson

Acknowledgments

I wish to thank the following companies and agencies that have all contributed in some way to this book: Shell Oil Company, General Motors Corporation, Ford Motor Company, McDonnell Douglas Corporation, The American Petroleum Institute, U.S. Department of Energy, U.S. Department of Transportation, U.S. Environmental Protection Agency.

I'm also grateful to Dave Malcheski, head of the Department of Energy's Driver Awareness Project in Las Vegas, Nevada for his helpful comments and suggestions.

Contents

Introduction

Unfortunately, there is no panacea for the current high price of gas, but HOW TO GET MORE MILES PER GALLON is a step in the right direction. It's a book for *all* drivers. By following even a few of the suggestions here you can dramatically increase your gas mileage. Each paragraph reveals a specific mileage aid—some will save you gallons of gasoline, others, only a drop or two, but you are sure to save money in the long run.

The United States with only 6% of the world's population uses over 35% of the world's energy. This alarming consumption has made us heavily dependent upon foreign oil sources to meet our demands. We currently import approximately 50% of the total petroleum we use. Transportation is the single largest user of petroleum in the country, accounting for 50% of the total consumed, or the equivalent of the entire amount of oil we import!

The average driver gets less than 13.7 mpg from his car and uses somewhere between 600–700 gallons of gasoline per year. The 110,000,000 registered automobiles, 30,000,000 registered trucks and 500,000 buses consume a staggering 300,000,000 gallons of gasoline and diesel fuel *each and every day*. If the average automobile fuel economy could be improved by only 15%, we would cut 28,000,000 gallons per day from that total. Another 8,400,000 gallons daily could be subtracted if every driver who now ignores the 55 mph speed limit were to obey it. If we were to add only one person to the average commuter passenger load, an additional savings of 35,000,000 gallons per day would become a reality. These are goals that can be achieved right now.

The mileage improvement you get from your car will be directly related to how closely you follow the gas saving tips in this book. By minimizing rolling resistance and aerodynamic drag, increasing engine efficiency and using economy driving techniques, Shell Oil Company Mileage Marathon drivers recently coaxed a car almost 400 miles on a single gallon of gas! These three factors, along with many others, are discussed here in non-technical language. Driving and parking techniques; how, when and where to buy gasoline; practical and workable gas saving additions and options; fuel conserving alterations, adjustments and inspections—these all play vital roles in improving gas mileage. But you, the driver, are the single most important factor. Only you can make the conscious decision to drive economically. By using this book as a guide you'll be well on your way to becoming an expert economy driver. In these days of skyrocketing gasoline prices and the ever—present threat of another gas shortage, it is comforting to know you're squeezing every possible mile from each gallon. The effort you make will not only increase gas mileage, conserve fuel, save money and minimize car repair bills, it will lessen pollution, save lives and speed America on its way to energy self-sufficiency.

How to Use This Book

At first glance nearly 300 ways to save gasoline may seem a bit overwhelming, but you will soon see how one suggestion readily relates to another. Each tip pinpoints a specific mileage improver, whether it is a gas saving addition, an economy driving method or an easy do-it-yourself adjustment.

Read through the book once from beginning to end to get a feeling for its scope. Then go back and mark the parts that are of special interest to you. These suggestions will make up your basic program, your starting point in the quest for better mileage. Incorporate them into your driving and they will soon become habits.

Keep the book handy—the glove compartment of your car is an ideal spot—and refer to it often. Browse through a few pages while parked or when confronted with unavoidable delays. Pick out a couple of techniques and practice them as soon as you get moving, while they're still fresh in your mind.

Commit yourself to fuel conservation and be aware of your driving at all times. Write the words "THINK ECONOMY" on a piece of paper and tape it to your dashboard. This will act as a constant reminder.

Check your gas mileage with each fillup. As you see the mpgs increase, you will be encouraged to add even more conservation measures to your driving regime.

For additional information about better gas mileage and an illustrated brochure of many of the products and items mentioned in this book please write to:

THE BETTER MILEAGE COMPANY
P.O. Box 40063
Tucson, Arizona 85717

Driving Techniques that Will Save You Gas

On the Highway

Slow down! Keep your top speed under 55 mph if you want good gas mileage. Excessive speed will put holes in your gas-money pocket. Most cars give best mileage at about 35 to 40 mph. In this speed range, engine efficiency is at maximum and wind and rolling resistances are relatively negligible. For every mph over 40, wind resistance increases proportionately and gas economy suffers. Have you ever had a 60 mph wind blowing in your face? It would knock you down. This is excactly what a moving vehicle must overcome, because a car traveling at 60 mph in *still* air is encountering the wind-resistance effect of a 60 mph headwind on a motionless car! Depending on car size and frontal area, it takes 10–15 percent more gas to travel at 45 mph than at 35 mph, and another 10–15 percent to speed up from 45 mph to 55 mph. Above 55 mph, mileage suffers even more severe penalties. A car moving at 70 mph gets only two-thirds the gas economy of one going 45 mph—a heavy penalty to pay for excessive speed. To further dramatize the effects of speed and wind resistance, it has been demonstrated that a large van with approximately 100 square feet frontal area requires over 100 extra horsepower to overcome wind resistance at 70 mph, as compared with a van traveling at 50 mph. Figure 1 illus-

Figure 1. Road Load Fuel Consumption

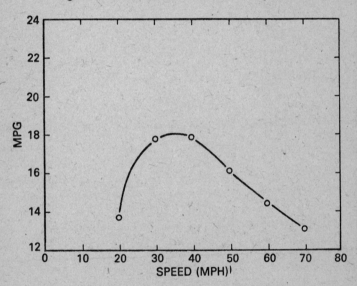

trates speed vs. gas-mileage. The wind resistance, or aerodynamic drag, depends upon the speed and shape of the vehicle. The more streamlined and the less frontal area, the less aerodynamic drag. This aerodynamic drag is represented in a factor called "coefficient of drag" or

C_D. The lower the number, the more efficient the car is at higher speeds. Figure 2 gives the C_D of different-shaped cars.

Figure 2. Air Drag (C_D) of Typical Full-Sized Cars

	PERIOD	C_D (APPROX.)
	Late 1920's Early 1930's	0.70
	Late 1930's Early 1940's	0.58
	Late 1940's Early 1950's	0.52
	Late 1950's Early 1960's	0.50
	Present	0.47

Courtesy: McDonnell Douglas Corp.

Obey the national 55 mph speed limit and you will save gas and contribute to highway safety at the same time. The speed limit was made law so we could preserve our nation's dwindling petroleum reserves; obey it, and you'll do your country and yourself a service.

Drive with the windows closed whenever you can. Open windows, especially at high speeds, create more wind turbulence and have the effect of "holding back" the car. It takes extra gas to overcome wind drag; in fact, at highway speeds, open windows can lower mileage by as much as 10 percent.

When driving on a two-lane highway with a slow-moving vehicle ahead, don't play hide-and-seek behind it. Darting in and out looking for a chance to pass destroys momentum, uses brakes unnecessarily, and plays havoc with mileage. Drive ahead. If you see a slow-moving car, ease off the accelerator and coast to within a safe distance of it. It's easier and safer to check for oncoming traffic when you keep a respectable distance behind and the view isn't obstructed. Once the way is clear, apply steady pressure to the gas feed and pass. It's cheaper and safer to pass this way; you'll also avoid a great deal of wear and tear on your nerves.

Recent tests involving long-haul truck drivers have shown a definite relationship between driving fatigue and poorer gas mileage. Make sure you get plenty of rest—don't drive past your physical capabilities and you will drive safely and economically.

When on a long trip, drive in your stocking feet occasionally. This gives you a better feel of the accelerator pedal and keeps you alert.

A very valuable method used by economy-minded drivers to better highway mileage is to accelerate slowly to a point a few miles under your desired speed, then e-a-s-e off the gas until the car reaches that speed, and hold it there with steady pressure. If you have a vacuum gauge you will see the needle rise, indicating better economy, as the car is eased into its cruising speed. In racing language this is known as maintaining speed with minimum throttle.

City Driving

Figure 3. Fuel Costs, In Dollars, Per 10,000 Miles

Example: If you pay an average of $1.15 per gallon and your car get 12 MPG, your fuel cost for 10,000 miles of driving is $958. If you drive 20,000 miles a year, your annual fuel cost will be twice this figure, or $1916. If you own a car that gets 20 MPG, your fuel cost for 10,000 miles at $1.15 per gallon is $575.

Combined City/Highway MPG

Cents Per Gallon

MPG	$ 1.30	$ 1.25	$ 1.20	$ 1.15	$ 1.10	$ 1.05	$ 1.00
50	$ 260	$ 250	$ 240	$ 230	$ 220	$ 210	$ 200
48	271	260	250	240	229	219	208
46	283	272	261	250	239	228	217
44	295	284	273	261	250	239	227
42	309	298	286	274	262	250	238
40	325	312	300	287	275	262	250
38	342	329	316	303	289	276	263
36	361	347	333	319	305	292	278
34	382	368	353	338	323	309	294
32	406	391	375	359	344	328	312
30	433	417	400	383	367	350	333
28	464	446	428	411	393	375	357
26	500	481	461	442	423	404	385
24	542	521	500	479	458	437	417
22	591	568	545	523	500	477	454
20	650	625	600	575	550	525	500
18	722	694	667	639	611	583	555
16	812	781	750	719	687	656	625
14	928	893	857	821	786	750	714
12	1083	1042	1000	958	917	875	833
10	1300	1250	1200	1150	1100	1050	1000

*Figures updated October, 1979

The Environmental Protection Agency says that travel habits in the U.S. lean heavily toward driving conditions which give poor fuel economy. U.S. autos accumulate about 15 percent of their mileage in trips of 5 miles or less; however, these trips consume more than 30 percent of the nation's automotive fuel, because autos operate so inefficiently in short trips. Let's turn now to city-driving techniques and see what we can do to cut into that 30 percent figure.

Any car momentum (forward movement) you can conserve is better than none. It takes up to 6 times as much gas to get a car from a dead stop than from a moving speed of just a few miles per hour. Cutting down on the number of times you have to stop is a major requisite for becoming a good economy driver. By eliminating many complete stops you gain from 10 to 25 percent better gas mileage. Preserving momentum is an absolute must and is especially important for city driving. Your gas dollars pay for your car's momentum, so don't waste it—any maneuver that conserves momentum saves gas.

Every state has "School Zones" with posted speed limits requiring a driver to slow down and stop for children. Avoid these when you can and eliminate yet another "gas trap."

If a traffic light turns red as you approach it, and you know that it will be red when you reach it, ease off the gas and coast up to the light as *slowly* as possible. Even though you will have to stop, a slow, coasting approach uses the gas carrying you to the light that would have otherwise been wasted idling at the intersection. The slower you approach, the more gas you will save. Use this method any time you have a mandatory stop ahead.

Freeway rush-hour traffic can be likened to an accordian, stretching out and compressing over and over again. Cars creep along at a snail's pace, stop, speed up a bit, stop again, and so on. This continues until traffic thins and normal speeds are resumed at the city's outer limits. Try to keep your speed steady—no matter how slow—and avoid as many stops as possible by keeping some distance between yourself and the car ahead. This will act as a cushion and give you room to coast. Remember, any momentum—even as little as 2–3 mph—is much better than having to start from a dead stop. Turn off the engine or take the car out of gear during long, unavoidable delays.

According to a recent EPA *Emissions and Fuel Economy Report*, it takes 50 percent more gasoline to drive under urban conditions than on open highways. This is why mileage figures published for various new cars often show dramatic differences between city and highway fuel economy. To obtain maximum mileage in city conditions you *must* know the timing of the traffic lights. There are probably many lights in your own town that you pass countless times each day. Why be trapped by these lights day after day and let them rob you of mileage you should be getting? With a little effort you can find out how long the lights stay red or green. Glance at the second hand on your watch or count slowly to yourself as one changes. Most lights operate in the fifteen-second to one-minute range, although some may be longer. By knowing how long a light stays red or green you can slow or increase your speed accordingly.

Say you are a block from an intersection and you see the light turn red. From previously having timed the light you know it stays red for fifteen seconds; you therefore immediately slow down a bit and pace yourself so that the light is green by the time you arrive at the intersection. No lost momentum here. Practice this and soon you will find yourself timing lights everywhere, making fewer stops. You'll be able to use this method countless times, each time deriving satisfaction from the knowledge that you just saved a good deal of gas. By timing lights you get to where you're going just as fast as the guy who guns his motor at the green only to be stopped a block or two ahead by another red. The drivers in the old Mobilgas Economy Run realized how critical stoplights and traffic patterns were to good mileage and would spend days timing lights and observing traffic along the proposed city section of the route. Then, on the day of the Economy Run, they could zip through the city without stopping and gain those extra mpg's needed to win.

Just as avoiding stoplights saves gas, eliminating turns will do likewise. Drive "as the crow flies" for best economy. If two routes are the same in every respect except that one has more turns, the route with the least number of turns will be the most economical. Once the car is moving, its most effortless direction is straight ahead. Plan your trips about town to eliminate as many turns as possible. Each time you turn, valuable momentum is lost that can be regained only by burning more gas.

An economical driver is a skilled driver. If you have problems parking or maneuvering in close quarters, practice to overcome them. It will pay off in a lifetime of gas savings and peace of mind.

Group as many errands and stops into a single trip as you can. It's better to do more during one trip while the car is warm than to make one or two extra trips with a cooler engine. Consolidate your shopping when possible.

As you approach an intersection to make a left turn, try to time your approach so that all of the oncoming traffic has passed by the time you begin your turn. In many cases, this will eliminate an unnecessary stop.

Although city driving consumes much more gas than highway driving, it does offer you one way to drive at maximum economy potential, something which highway driving can't do. The trick is simply this: Maintain speeds as close as possible to the economical 30-40 mph range. Obviously, on the open highway you can't drive at 30 mph, but in the city there are countless times when you can. Take advantage of these situations and keep your car moving in the economy range whenever possible. Although it seems to contradict logic, it is nevertheless true that it takes significantly less gas to travel at 30–40 mph than it does at 20 mph. Get your car up and over the 20 mph mark when you can. Each mph closer to 30 means increased gas economy.

Take advantage of the metal crossbars placed in the road near many traffic-light intersections. When activated by the weight of your passing car, the bars (sometimes they are hidden beneath the road surface) trigger a mechanism that causes the light to change from red to green. Know where these are located and you can slow down as you cross them, allow the light to change to green while you are still moving, and continue on through the intersection without stopping. It's another effective way to conserve forward momentum and save gas.

On four-lane (two each way) streets, where an additional lane is provided at intersections for left turns, the best bet for stop-free driving is usually the one closest to the center or passing lane. The left-turn lanes at intersections funnel off left-turn traffic and keep the center lanes moving at a steadier pace than the curb lane, which must slow down or stop for buses, parking cars, taxis, passenger pickup, and right-turn traffic. In city rush-hour madness the center lane will get you home faster and cheaper.

Here's a gas-saving driving technique you'll benefit from a thousand times each year. When approaching a stop sign where there are cars already stopped, gauge your speed so that the cars have passed through the sign by the time you arrive. By doing this you will have to stop only once, instead of stopping, pulling up as one car leaves, stopping a second time, pulling up again as the second car leaves, and so on. Slow stop-and-go eats gas, and this is one excellent way to beat it.

When you see a traffic light turn green a block or so ahead and you know (because you now time the length of the lights, as every economy-minded driver should do) it will turn red at about the time you arrive at the intersection, prudent application of the accelerator is in order. You will use less gas by making the green light than by stopping at the red and having to start over. Don't gun the car, and stay within specified speed limits or you could nullify any savings by getting a speeding ticket.

For better economy in stop-and-go traffic on extremely hot days, place the transmission in neutral whenever you are stopped. It eases the strain on the already over-worked transmission and prevents it from overheating. The engine will benefit also because it won't have the added burden of turning the idling transmission.

Have correct change ready *before* stopping at a toll booth or *before* leaving a pay parking lot. Have letters stamped and ready for mailing at curbside boxes. There are instances galore, similar to the above, where a little planning will cut down on engine-idling time. Remember: If the car isn't moving and the engine is running, you're getting 0 mpg.

If you are fortunate enough to live in a state where right turns on a red light are legal, take full advantage of this fuel-conserving law. As soon as the way is clear, make your turn—don't waste gas waiting for the light to turn green.

Don't drive into obvious "gas traps." If you want to make a turn at an upcoming intersection, but see a long line of cars already waiting, turn a block or two sooner and reroute around an obvious gas-wasting situation. How many times each day do you find yourself stuck at problem intersections? With a little foresight you can turn sooner, reroute for a few blocks, and avoid these irritating and costly traps. Remember to reroute when circumstances ahead dictate.

If yours is a two-car family—and almost every family is nowadays—use the car that is already warmed up when running about-town errands. There's no sense in warming up the second car and paying through the nose for cold-engine gas consumption when you can use the warm, more economical car. This is an especially valuable practice in freezing climates where an engine never becomes fully warmed on short trips. Also, in the already-warmed car the heater will function better. You will enhance the engine life of *both* cars by using the warm one. Figure 4 shows that with an outside temperature of 10 degrees Fahrenheit a car must travel at least

14 miles to attain maximum warmed-up city economy. On short trips of one or two miles a car may get *less than 10 percent* of the mileage it is capable of when fully warm.

Figure 4. Fully Warmed-Up Economy Versus Trip Length

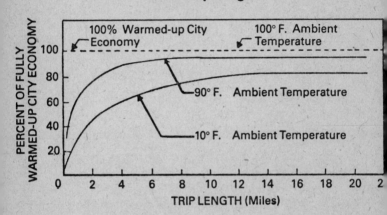

Drive ahead. Anticipate traffic situations. Be alert. All contribute to better mileage and automotive safety.

Economy Tricks for *All* Driving Conditions

Before getting into your car, look around. Check in front, in back, and to the sides of your vehicle. By doing this you get an advance scouting report on how easy or hard it will be to get the car moving, and you cut down

Figure 5. Cost to the Individual Motorist

Due to the fuel economy effects of trip length, a typical family car can take the following trips on 25 gallons of gas:

Ten 40-mile trips, or
Sixty 4-mile trips, or
Ninety 2-mile trips, or
One hundred 1-mile trips.

At 75¢ per gallon, the fuel costs would be:

4.7¢ per mile on 40-mile trips, or
7.8¢ per mile on 4-mile trips, or
10.4¢ per mile on 2-mile trips, or
18.8¢ per mile on 1-mile trips.

Courtesy: U.S. Environmental Protection Agency

on the time spent looking around after you are in the car and the engine is running.

Before you start your engine, make sure you are ready to go. Adjust seats, light cigarettes, put packages in the back seat, strap the kids in, and do everything you must do *before* you turn that key.

EPA tests have shown that it is more economical to turn the engine off rather than let it idle if the idle time exceeds 60 seconds. In other words, if the car is going to idle longer than 60 seconds, you save gas by turning the engine off and then restarting it when ready. If the engine is going to idle *less* than 60 seconds, let it be, for it takes more gas to restart it than a minute of idling will use. There are numerous occasions each day when gas could be saved this way: long stoplights, train crossings, freeway congestion, waiting in parking-lots or on drive-in bank lines, to name a few. You probably can think of many others.

On the average, an engine will use about a gallon of gas for ever 1–2 hours it idles. By using the above method you will accumulate good gas savings in short time. Remember the rule: Under 60 seconds, let it idle; over 60 seconds, turn it off.

If you're driving a car with a standard transmission, get into high gear as quickly as possible. On level roads you should be in high gear before the car reaches 20 mph. Low-range gears use much more gas than their higher counterparts. For example: At 20 mph, second gear uses as much as 15–20 percent more gas than high. First gear uses 30–50 percent more! Help your automatic get into the high range faster by easing off the gas just as the transmission reaches its shift point. Getting into high gear fast is one of the economy-run driver's "must" techniques. You can use it also, and enjoy the additional miles per tankful it provides.

Skip a gear when you can. Shift directly from low to high if conditions permit. Level or downhill starts are good times to skip second gear (or even start out in second gear, skipping first) and go directly to third. There are many opportunities, particularly with 4- or 5-speed transmissions, to avoid one or even two gears when shifting. You spend less time in the low, gas-consuming gears and more time in the economical higher ones.

Never shift to a lower gear for the sole purpose of slowing down the car. Except in very hilly country the brakes will do just fine and you won't have to pay the lower-gear mileage penalty.

"Jack rabbit" starts are murder on gas mileage. One or two fast starts with the accelerator floored will nullify gains made elsewhere. Take it e-a-s-y when you start out and push the accelerator down s-l-o-w-l-y. This is another "must" rule for top economy. Department of Transportation tests have proved that "jumpy starts and fast getaways can· burn over 50% more gasoline than normal acceleration."

To avoid hasty starts, envision an egg between your foot and the accelerator pedal. Now start out by pushing down in such a manner that the egg won't break. Another good trick is to pretend there is an apple on the front of the hood. Pull away in such a manner that the apple won't roll off. Use these mental tricks and they will soon form a good economy habit. Remember, no gas-saving device or practice can take the place of a light foot on the accelerator. It's the number-one consideration for top mileage. The lighter the foot, the better the mileage. You must obey this rule if you want top economy.

The levelest route with the least number of turns, the least amount of traffic, and the least number of required stops is the most economical.

Any time the engine is running and the car isn't moving, you're wasting gas.

Any time you are going down a grade, take full advantage of the fact that gravity is working in your favor and ease up on the accelerator. Let the car coast of its own volition, even if it means slowing down a bit. Downhill driving affords an excellent opportunity to bolster your gas mileage. A car going *up* a 4 percent grade at 50 mph gets about 9 mpg. That same car going *down* a 4 percent grade at 50 mph will get an astounding 35 mpg—a difference of 26 mpg. This is way more than enough to compensate for the gas used going up the hill. Use even the slightest downgrade to your advantage. Apply as little accelerator as necessary to keep the car moving at a reasonable speed and let gravity do the rest.

Reroute around time- and gas-consuming construction detours. If your day-to-day work route is being repaired, don't keep going back over it. Find an alternate smoother route with less stops and use it until road repairs are complete.

Don't tailgate! Much gas is wasted by driving too close to the car ahead. You must constantly brake and accelerate, brake and accelerate. By tailgating, you let the car in front dictate how you drive—and that is foolish economy. Stay a safe and reasonable distance behind the other car and you won't have to play the slow-and-go game. It's murder on mileage. Keep smooth, even accelerator pressure and pace yourself so that you are not continually slowing and stopping in cadence with the car in front but are keeping an air cushion between the two of you.

Drive into curves properly. Apply brakes and slow down *before* reaching the curve, then accelerate into and through it. In this way you are in complete control of the auto and get through the curve in the most efficient and safest way.

Know how to drive up a hill. When approaching a hill, try to get a "running start" at it. Build up as much momentum on the approach to the hill as is safely and legally possible. As the car begins to move up the hill, let the forward momentum carry it and use only as much accelerator as necessary. When you near the top of the hill, ease off the gas and let the car coast up and over the crest. Ideally, velocity should be maximum at the bottom and minimum at the top. On very long upgrades keep even accelerator pressure, again easing off as the top is reached.

Driving up hills is unavoidable and costly. A 4 percent upgrade can reduce car gas economy by 40 percent, so minimize the effect by using the above method. Figure 6 demonstrates the estimated effect of different grades on vehicles traveling at various speeds.

Once you have reached your desired speed, whether in the city or on the highway, keep it constant. You use less gas by going at a steady speed than by intermittently slowing and accelerating. *On the highway, varying speed by only 5 mph can reduce economy by as much as 1.3 mpg.*

"Drive ahead" and you see traffic situations developing and are better prepared to cope with them. You avoid needless slowing and stopping, and conserve critical momentum. "Driving ahead" is also equated with safety

Figure 6. Estimated Effect of Grades on Fuel Economy for a Typical Standard-Size Vehicle

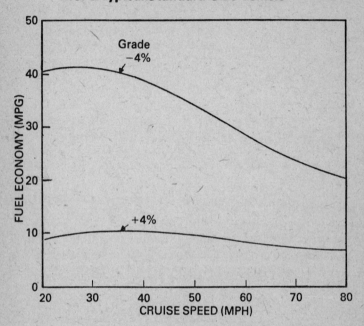

driving because you are more aware of the traffic in front and to the sides of your car and react to it more wisely.

When you must carry extra weight, try to *distribute it evenly throughout the car.* Packing heavy loads in the trunk will raise the front end, thereby increasing the frontal area. More frontal area exposed to the oncoming wind = more gas used. As speed increases, wind resistance increases, and the more frontal area the car has to "push" into the wind, the more gas it will use. In fact, a recent study by the Department of Transportation and the Environmental Protection Agency recommended a 10 percent decrease in the frontal area of all new cars as an effective and practical way to lower gas consumption. Spreading the additional weight throughout the car will keep it on a more even keel, minimizing additional wind resistance.

Use your rear and side-view mirrors often. Remember to "drive behind" as well as ahead and you'll have an advantage in the mileage game. Frequent glances in the mirrors will warn you of situations developing behind your car and you can adjust your driving accordingly. Besides being an obvious safety habit, using the mirrors frequently lets you avoid many turns and stops, thus saving momentum and gasoline.

After arriving at your destination, turn the engine off immediately! Needless idling wastes gas. However, there is one exception to this: When stopping after a continuous high-speed run, it is better to let the engine idle for a minute or so before turning it off. This helps eliminate "hot spots" and relieves hot fuel vapors that could cause vapor lock and hard starting afterwards.

Never rev the engine before shutting it off. Too many drivers have this habit. They erroneously think this extra "shot" will circulate oil for better protection when the engine is off. Actually, the opposite is true. A surge of raw gas floods the cylinders, doesn't have time to be ignited, dilutes the oil, and washes away vital cylinder coatings.

If you are low on gas and feel the engine cough and sputter (hopefully it won't happen to you), turn it off immediately. This keeps some fuel in the lines and the carburetor, and makes for much easier starting once fuel is replenished. You will avoid long periods of starter grinding and you won't waste gas with frantic accelerator pumping.

Almost forgotten these days is one of the most helpful built-in gas savers on your car—the emergency brake! Today's driver practically forgets it's there, and in doing so passes up many gas-saving opportunities. Using the emergency brake when you can saves wear on the clutch or transmission and conserves fuel to boot. Don't ride the clutch to keep your car at a standstill on hills but use the emergency brake instead. The same applies to cars with automatic transmissions. Holding the car on a hill by applying more gas only wastes fuel and makes the transmission work when it should be resting. Use the emergency brake and let the engine and transmission turn at normal speed.

Lower gears definitely consume much, much more gas than the higher ones; however, if the engine begins to lug and strain, shift to a lower gear fast, as *it* will be more economical. The engine uses more gas and works harder when it lugs in a high gear. A quick glance at the accelerator pedal will verify this. In high gear, with the engine laboring, the accelerator will be nearly floored and the throttle wide open, permitting full flow of gas. By shifting to a lower gear you get smoother engine response with the accelerator depressed only a fraction of what it was previously. Sure, you are in a lower gear, but you are saving gas. Don't be afraid to shift when the engine dictates it. Of course, with an automatic transmission the up-down shift cycle is taken care of "automatically."

Don't drive a car with one foot resting on the brake pedal. This is especially important on cars with power brakes, where the slightest pressure will partially engage the brakes. Don't force the car to fight itself—keep your foot off the brake.

In standard-transmission cars, don't rest your foot on the clutch pedal. A small amount of pressure can partially disengage the clutch, causing it to slip and reduce drive-train efficiency. Keep your left foot on the floor.

Avoid sudden stops. They can cause fuel to slosh out of the carburetor bowl or gas tank and can flood or stall the engine.

When stopped, take your foot off the accelerator and let the engine idle at normal rpm. Too often drivers have a tendency to "rest" their foot on the pedal, and as a result they waste gas. Make it a habit. When stopped, foot off.

You can "help" your automatic transmission by easing up slightly on the gas pedal just prior to feeling the transmission begin its shift. Easing the accelerator hastens and smooths the shift because of the increased engine vacuum created. This saves gas, too. With a little practice you will be able to "feel" the transmission start its shift and ease up on the gas accordingly. If you have a vacuum gauge, notice how the needle swings into the economy range as you practice this trick.

Before you start out, make certain the parking brake is fully released. Too often drivers let this silent thief rob them of mileage they should be getting.

Cars equipped with a dual-range transmission offer a choice of D1 or D2 driving. Always use the D1, or higher range, for maximum economy. D2 should only be used on the rarest occasions, as decelerating down a very steep hill where the transmission can help brake the car. At all other times use D1.

Use passing gear for emergency situations only.

Never pump the accelerator or race the engine while the car is standing still.

When starting a *warm* engine there usually is no need to push the accelerator pedal down prior to turning the key.

If you have tried everything and still have a problem with engine run-on or dieseling, try leaving the car in gear when turning the key off. This usually does the trick. Be sure to apply the brake pedal while doing this.

Think ECONOMY at all times. This is important. You must motivate yourself to be a successful economy driver. Keep economy and safety foremost in your mind.

Keep a steady hand on the wheel at all times. At first glance this may not seem too important, but it is very crucial if top mileage is to be had. Don't let the car wander from side to side. Hold it steady and on a straight course. Any side-to-side movement detracts from forward momentum and can cost you 1–2 mpg at highway speeds.

Cold Starts—That Critical First Minute

If the engine floods while you are trying to start it—strong gas odor and wetness around the carburetor are usual indications—don't continue to pump the accelerator. This wastes gas and makes the car even harder to start. Instead, push the accelerator pedal to the floor and hold it there while turning the starter. This opens the throttle and allows the excess gas to drain out of the carburetor. A minute or two of cranking the starter *while holding the gas pedal down* will usually get it started. If the engine is equipped with an automatic choke, the butterfly valve must be fully opened while the starter is turned. If the engine is cold, more than likely the valve will be closed. Have someone hold it. or place an object in such a way as to hold the valve open until the engine starts. Make sure the object can't be drawn into the carburetor.

Push the accelerator completely to the floor when starting a cold engine. This will activate the automatic choke mechanism and close the carburetor butterfly valve. One pump, or two at the most, should be sufficient to start the car on even the coldest mornings. More pumping only wastes gas and is probably an indication that the choke isn't adjusted properly.

After a cold engine has run a few mintues it is a good idea to tap the accelerator pedal lightly to make sure the high-idle cam has been disengaged. You should notice a considerable drop in engine-idle speed with this easy maneuver.

Figure 7. Average Fuel Economy in City Driving at 10° F. Beginning with a Cold Start

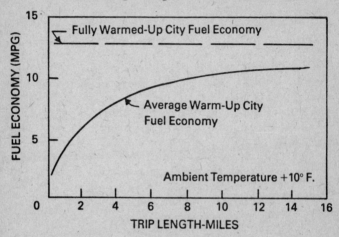

Years ago it was considered a good practice to let the car idle a minute or two after cold-engine starts. Today, with higher-compression engines, the opposite is true. After you start your car get moving *immediately.* An engine under road load conditions will warm faster and lubricate more efficiently than one that is idling. Use slow speeds for the first mile or two, then increase to cruising as the engine gradually warms. Economy improves with distance traveled. Remember to take it easy that first mile or two, as that's the time gas consumption is at its peak. Look at Figure 7 to see how fuel economy suffers during the first few miles after a cold-engine start. On a cold day (10°F) a car may never reach its full economy potential in city driving.

Not only does fuel economy suffer greatly immediately after a cold start, but most engine wear occurs then. In fact, McDonnell Douglas engineers have recently demonstrated that *90–95% of all engine wear occurs during the first ten seconds after a cold start!*

The Emergency Economy Method—How to Go 30–100 Miles on Only 1 Gallon of Gas

It's late at night and you're driving along a deserted stretch of road with the wife and kids. You're very low on gas because the last station you stopped at was closed, but the map shows a small town ahead and you decide to keep going and fill up there. When you reach the "town"—panic! Nothing there, just a few deserted buildings and a railroad crossing. Frantically you look at the map—the next town is 50 miles away! You estimate about a gallon of gas left in the tank, and you know that your car gets around 20 mpg on the highway. A gallon will never take you the 50 miles. Wrong!!! Not if you know the Emergency Economy Method. That gallon will get you to the next town—and you may even have a little left over! How do you do it? You simply use the coast-accelerate-coast technique of the Mileage Marathon drivers.

Accelerate s-l-o-w-l-y to 20 mph, then quickly turn off the ignition and shift the car into neutral. Let the car slow to 5–8 mph, start the engine, and repeat the process. That's all there is to it. Repeat this procedure over and over again and you'll be able to double and even triple normal gas mileage. Remember the technique: Slowly accelerate to 20 mph. Turn the engine off and let the car slow to 5-8 mph. Start the engine and repeat the process. Simple, yes—but incredibly effective.

Although this method is obviously impractical for everyday use, it's comforting to know that in the event you are ever caught in an emergency, you can coax many extra miles from your car. We feel it's better to be safe and secure in your car going 5–20 mph than to be stranded and out of gas. If you have to use this method only once in your lifetime, you'll be glad you knew it. (NOTE: Do not use this method if your steering wheel locks when the ignition is turned off or when you are driving down steep grades. Also, consult local and state vehicle laws before using the Emergency Method, as coasting, even in emergency situations, may be illegal in your area.)

The telephone can save you gas??!! Definitely! Simply do as the man says and "let your fingers do the walking." Use your phone book and phone *before* starting out to a place of business. Make sure that it is open and that the item you want is in stock. Many gas-wasting trips can be avoided by using the phone first.

2 Parking Techniques that Will Save You Gas

It's much easier and more economical to do difficult maneuvers such as parking when the engine is warm rather than cold. Steering gears, wheel bearings, transmission, and differential also work better when their respective parts are warm. It's good sense when leaving your car overnight to park it in such a way as to make leaving in the morning easy. Do all backing, turning, and other maneuvering while the engine is warm and using less gas. You'll save a lot with this one technique because you ease the power requirements of the cold engine.

When parking in a lot that has double-row parking (most lots in shopping centers, stadiums, racetracks, and such are arranged in this manner), choose a spot in the forward row. That way, you won't have to back out when you leave, so you're eliminating an unnecessary maneuver.

When you're parked in a street that has a traffic light at the intersection ahead of you, watch the light and gauge

27

your start to coincide with the green. Don't start the engine until the light turns green. Once it changes, pull out into the traffic flow and through the light. You save gas because your engine doesn't idle needlessly, and because you don't have to start from a dead stop twice (you would have had to if you pulled out while the light was red).

Choose the parking space closest to the intersection if possible. It's easier to get in and out of, and you won't waste gas jockeying the car around as you would in a tight spot.

If no end space is available, try parking so there is plenty of room between you and the car in front. This will make leaving a lot easier and will save gas. A good rule of thumb is to park in a manner that will make leaving the easiest.

Before backing into or pulling out of a tight parking place, turn off all accessories. The engine is already laboring overtime turning the power steering unit, and any added accessory pull foolishly wastes gas. When leaving, wait a few seconds until you are clear and moving before you use accessories—a small sacrifice that pays nice dividends.

When you have a full tank of gas, park the car facing downhill if on a grade to prevent any gas from spilling out of the tank.

Keep your car garaged or under a carport. If this isn't possible, then any type of shelter over or around the car will help. Trees, a wall, the side of a house—all provide some protection from the elements. In cold, windy climates shelter is a must, because moving air can greatly hasten the rate at which an engine cools. Shelter from wind and cold will do more to promote good winter gas mileage than any other single item. Remember, it took gasoline to warm the engine, and anything you can do to conserve the heat will improve mileage.

In hot, dry, or dusty climates a garaged car is impervious to gasoline evaporation. As much as a quart can be lost on extremely hot, windy days if the car is left outside. Garaging your car also eliminates the problem of dust entering the engine compartment where it can clog the carburetor and air cleaner. In wet climates, shelter guards against the frustrations of a wet ignition.

Parking in shady areas helps prevent gasoline evaporation that would otherwise occur if you parked in the hot sun. This can be crucial in some southern and western states, where the summertime temperatures can reach 120° and a carburetor bowl of gas can evaporate in a day. Your car will stay cooler, too, in shady areas, and that means less work for the air conditioner once the engine is started.

There are many instances each day when a driver can save gas by just parking a little sooner and walking a little further. Why cruise through parking lots time and time again trying to find a spot "up front" when there are plenty of empty spaces in the back? Take the first available space you see and don't be afraid to walk the extra 50 yards or so. If everyone practiced just this one item, the nation could save *billions of gallons* of gas each year. Slow stop-and-go driving is the most gas-consuming, so be willing to walk a little—you'll save a lot.

Before buying a new car, know its mileage rating. Ask your dealer for a copy of the current *Gas Mileage Guide* or write to Gas Mileage Guide, Pueblo, Colorado 81009.

Ways to Save at the Gas Station

Figure Your Gas Mileage Simply and Accurately

Here's how to figure your gas mileage. Jot down the speedometer reading to the nearest mile whenever you fill the gas tank. Let's say it reads 8,510. At your next fill-up, mark down the mileage reading again and also record the number of gallons of gas purchased. We'll use 8,760 and 10.2 gallons. Be sure to use the fill-up technique described on page 33 to get accurate gallons-used readings. Now we subtract our previous speedometer reading (8,510) from the current one (8,760) and get 250 miles traveled. If the speedometer has a correction factor as described below, be sure to add or subtract it from the miles traveled. Let's say our correction factor is +2 for every 100 miles traveled. This gives us 5 extra miles, which we add to the 250 to get 255 as our true mileage figure. Now divide that figure by the number of gallons used: 255/10.2 = 25 mpg. Twenty-five mpg is our gas mileage for that fill-up. Check mileage with each fill-up. This will give you an accurate running record of how you are doing as an economy driver. As your mpg's increase with each fill-up, you'll become encouraged to try more and more economy techniques or additions. Think economy at all times and let's begin conserving our nation's fuel—and our money, too!

If your speedometer isn't accurate, you may be getting better (or worse) gas mileage than you think. There is no need to take it to a speedometer shop to have it checked because you can do it yourself. Here's an easy, foolproof way:

Take your car out on the interstate or any smooth highway *where mileage markers are posted*. Note your speedometer reading at the exact time the car reaches one of the markers and write it down along with the number on the mileage marker. Try to read the odometer to the nearest tenth of a mile. For instance, let's say our reading was 21,570.3. Now drive 10 markers (10 true miles) and check the reading again as your car passes the tenth marker. For our purpose let's use 21,580.1. Now subtract the first figure from the second one: 21,580.1 − 21,570.3 = 9.8 miles. We actually traveled 10 true miles, as verified by the mileage markers, but the speedometer recorded only 9.8 miles; therefore, we're getting *better* mileage than the reading indicates. We must add two-tenths of a mile (9.8 + .2 = 10) for every ten speedometer miles we travel, or 2 miles for every 100. Be sure to add or subtract this difference when figuring gas mileage. If you traveled 350 miles on a tank of gas you would add 2 miles for each 100 traveled—in this case + 7 miles. So the true corrected mileage is 350 + 7, or 357 miles. This is the number to use when figuring gas mileage.

Of course, it can work to your detriment also. If the

As the day wears on, the gas will expand in your tank and you gain the extra expanded volume. If you bought the same amount of gas in the heat of the day it would have already expanded, and as the evening approached and the gas cooled, you'd lose volume, because gas seeks its cooler, more dense state. Try early-morning or late-evening fill-ups, and profit from the "extra" gas you receive at no additional cost.

When recording gas mileage, make a note of the brand of gas purchased at each fill-up. If one brand consistently gives better mileage than the others—and is competitively priced—it makes good sense to use it. Sometimes brands of gasoline *can* make a difference, so compare and save.

Keep your gas tank on the full side to prevent gasoline loss from evaporation and condensation. This practice can be especially helpful in the winter when condensate from a near-empty tank could freeze, blocking gas lines and making starting impossible.

If you frequent one of the popular self-service gas stations, don't be embarrassed to lift the hose at the end of gas delivery and wait for all the gas to run out. You can gain an additional half cup by this simple practice. You wouldn't think of leaving an item on the checkout counter of a supermarket, so why do it at a gas station? You need every edge you can get in the battle for better mileage.

Although gas "wars" are probably a thing of the past, you can still keep your eyes open for low-priced gasoline. Service stations a block or two from freeway off-ramps may offer the same brand of gas at 10¢ a gal-

odometer records more miles than you actually trave
then the extra mileage must be subtracted from the tota
miles to find your true mileage.

Fill the Tank Right

Insist on a *slow* fill-up if an attendant waits on you. Gas
that spills down the side of your car because of an over-
anxious attendant still costs you, but you don't get to use
it.

How many times have you noticed gas dripping from
the tank of the car ahead of you? It's easy to lose a
half-gallon of gas by filling the tank to the brim. Every
time the car starts or stops, gas sloshes out of the filler
neck. If the day is hot, the gas expands and forces itself
out. Don't "top off" the tank, but keep the gas level a
comfortable margin below the cap.

A good way to fill your tank without spilling and wasting
gas is to set the automatic shut-off at the *slowest* deliver
rate. Place the nozzle as far into the filler neck as it wil
go, and when the pump clicks off, round out the
amount to the nearest tenth of a gallon. This leaves the
tank a gallon or so from full and avoids gas spills. It also
gives you an accurate method of determining the exact
number of gallons used when figuring gas mileage. Try
to use the same pump and park the car in the same direction to get an even more accurate reading.

Buy your gas in the early morning when possible. In
cool of the morning, gas is more dense because the
hasn't had time to heat and expand it. Gasoline pu
measure *volume*, and although you receive only
amount the pump indicates, it is more "concentr

lon cheaper than their freeway counterparts. It pays to drive a bit and look around. It takes but a minute or two of your time and you could end up by saving a couple of dollars on a fill-up.

Whenever you take a long trip (one that requires at least one fill-up on the road) and plan to return by the same route, write down the names of the towns and filling stations that offer the lowest gas prices. Then, on your return, take advantage of these "gasoline oases" by timing fill-ups to coincide with the respective towns. Any traveler knows how gas prices can vary from town to town, so follow this easy method and pocket the extra savings.

Here's a tip that can save you many gasoline dollars when driving in a foreign country. Be sure—stand there and watch—that the station attendant resets the gas pump meter to zero ($00.00) *before* putting gas in your car. Literally thousands of dollars have been fleeced from unsuspecting tourists who were charged the amount of the *previous* sale *plus* the amount it took to fill their tank. In the U.S., most pumps are automatically zeroed or must be zeroed by hand before delivery can begin, but in many foreign countries this isn't the case. Get out of the car and make sure the attendant zeroes the pump before he begins filling your tank. We learned this the hard way many years ago in Mexico, paying for 120 litres of gas when the tank could only hold 80. (Some "country" stations in the U.S. may still have the old pumps, so it's better to get out of the car and make sure, if you have doubts.)

CAUTION! Never use leaded fuel in a car meant for "unleaded fuel only." Even if you got better mileage, you would ruin the catalytic converter in the process. What's more, it's against the law.

Octane—Don't Buy More than You Need

Don't pay for octane your car doesn't need. It's money down the drain. Plying your car with 70¢ premium when it will run as well on 60¢ regular is like the guy who takes huge quantities of vitamins each day, only to have his body reject what it doesn't use. The body knows how much of each vitamin it needs and uses only that amount. The same applies to your car, so use the lowest octane that provides good performance. You can save upward of $2 on each tankful by knowing the exact octane requirements of your car.

If you live in an area where octane selection is limited to regular and ethyl, here's an easy and economical way to custom-blend the exact octane your car needs. If the car doesn't need ethyl but seems to run a little rough (and knock) on regular, the correct octane lies somewhere between the two. Next trip to the gas station custom-grade your own. Fill the tank two-thirds full with regular and then top off the remaining third with ethyl. Experiment with each tankful, adding or subtracting a little ethyl until you find the exact "mix" your car runs best on. Mark down the amount of regular and ethyl used and then figure the ratio of the two. Let's say you bought 10 gallons of regular and 5 gallons of ethyl; the ratio then is 10 regular to 5 ethyl, or 2:1. So every time you fill up, use the 2:1 ratio and pay only for the octane needed. The lowest octane that gives smooth performance is the best buy.

If your car uses premium gasoline, you can save money

by purchasing regular whenever you take a trip over mostly level roads. At highway or cruising speed, engine load is only a fraction of horsepower available, and a car that must use ethyl under city driving conditions may do surprisingly well on regular. Try a tank of regular, or a mix of one-half regular and one-half premium, and you'll be pleasantly rewarded with equal response at less cost.

A car that must use premium gasoline at sea level will probably run just as well on regular at high elevations (5,000 and above). The reason is that as altitude increases, the car's octane requirements are lessened due to the change in the carburetor air/fuel mixture. Thinner air at high elevations means a *richer* air/fuel ratio, and *less* octane is needed to fire it. If you are going to be in the high country awhile, try switching to regular and save the 5¢–10¢ difference per gallon.

Many newer-model cars that use unleaded gas are experiencing engine knock, rough idling, and "run-on" (the engine continues to run after the ignition has been turned off). Switching to another *brand* of unleaded will sometimes cure these problems. Some brands have higher octane ratings than others, and a few numbers higher may be just what the doctor ordered to cure the problem.

Some Facts You Should Know About Octane

What is octane? Simply stated, it is the measure of the gasoline's ability to resist engine knock.

What does the $\frac{R + M}{2}$ number on the gas pump mean?

This is the Research Octane Number plus the Motor Octane Number divided by 2. The resultant number is commonly known as the $\frac{(R + M)}{2}$ antiknock index. It is the number you see posted on all pumps and the one we are concerned with.

How many times have you heard someone say that he owns such-and-such model car and it runs great on regular gasoline even though the owner's manual says to use premium? Yet you own the same model car with the same engine and it won't even start, let alone run, on regular. Is the first guy just pulling your leg? Probably not. Due to assembly-line variances in engine tolerances, octane requirements for the identical new car can vary by as much as 10 numbers! His car could be doing very well on 86 octane while yours could require a number quite higher.

Another thing to remember about octane: As your new car accumulates mileage, the octane requirement continues to *increase* until combustion deposits (carbon) stabilize. Octane requirement usually levels off at about 5 numbers higher than when the car was brand new. This means that an older car will always require a higher octane than when it was new.

Gas-Saving Items for Your Car

Inexpensive Fuel-Conserving Options

The *best* way to "learn" economy driving is to install a vacuum gauge in your car and let it be your teacher. The gauge is clearly marked to indicate when you are driving economically and when you are wasting gas. It monitors engine vacuum, which varies with the amount of pressure you apply to the accelerator pedal. If you watch the gauge you can't help but improve your gas mileage. A vacuum gauge can be purchased for as little as $5 and is simple to install. You also get a valuable added bonus, because a vacuum gauge can detect a whole range of engine ailments before they become serious. By correcting these immediately you save yourself costly bills later. If you buy only one add-on piece of equipment for your car, make it a vacuum gauge.

Water vapor, or water-alcohol vapor injectors, are one of the "gas-saving" commercial devices that really work. Ever notice how your car seems to run better on rainy days? Moisture in the air creates a more even-burning fuel mixture and the car seems to respond with smoother performance. Vapor injectors use the same principle: Add moisture and get increased volatility from the air/fuel mixture. The U.S. Government during World War II experimented with water vapor injection and today it is still being considered as a method of conserving our rapidly dwindling fuel supply. A water-methyl alcohol mixture, when injected into an internal combustion engine, *will* improve performance and increase gas mileage, but one must consider whether the price of the alcohol is offset by the increased gasoline savings. Many brands of water-alcohol injectors are now on the market at prices from $20 to $40. They all employ the same basic principle: Additional outside air is drawn through a container of water-methyl alcohol, causing the mixture to bubble. The mist from the bursting bubbles is then drawn into the intake manifold at some point below the carburetor and sucked into the engine. As the water-methyl alcohol mist combines with the normal carburetor mixture it produces a better-burning, cleaner fuel, thus enhancing pep and fuel economy. It may be worthwhile to experiment with vapor injection on *your* car.

Consider the price of the injector unit and the cost of a container of methyl alcohol (a half gallon usually lasts from 4,000 to 7,000 miles). Weigh these figures against any increase in miles per gallon you get, and then decide for yourself if that increase will eventually offset the money spent for the unit and the cost of additional methyl-alcohol refills.

In lieu of, or in combination with, a vacuum gauge, a tachometer is a helpful instrument for improving driving technique and gas mileage. The "tach" will indicate engine rpm's, giving the driver the correct and most efficient point at which to shift gears. You avoid lingering in the gas-consuming lower gears and eliminate costly and damaging engine "lug" by shifting at the proper times.

A fuel-consumption meter (fuel-flow meter) is a device that measures the exact miles per gallon your car is delivering at any time or for any given distance. It is accurate to 1/100th of a gallon. For the true gas-mileage enthusiast it is a must because you can actually see what mileage you are getting at any particular moment. The only drawback it has is price. Cost per unit is in the $60 to $80 range. This meter and a vacuum gauge make for probably the best combination of gas-saving devices available today.

Some cars come equipped with a magnetic oil-drain plug that attracts loose metal particles suspended in the oil. If your current auto doesn't have one (chances are it doesn't), it can be purchased at many auto accessory stores, or if you like, simply attach a small magnet to the inside of your present drain plug. By attracting oil-suspended metal particles, the magnetic plug reduces abrasion between moving parts, thereby lessening internal friction, preventing engine wear, and promoting economy. Be sure to clean the particles off the plug at each oil change—you'll be surprised at how many the magnet attracts.

Spending a few dollars on a locking gas cap could save you from being siphoned to the tune of a $15 tank of gas. It also protects you against pranksters throwing dirt or foreign objects into the gas tank. If you do not have a locking gas cap on your car, purchase one as soon as possible. Use the gas you paid for—don't let your car be an open invitation to gas thieves.

A few late-model cars feature as standard equipment a speedometer buzzer that can be set to go off at a pre-selected speed. During monotonous highway cruising, drivers have a tendency to speed up as the accelerator foot becomes heavier. The buzzer is an excellent way to prevent this from happening because it alerts the driver and warns him to reduce speed. It should be standard equipment on all cars. It's a gas saver and an excellent safety device. Some auto supply stores have the buzzers at modest cost, but installation may require some skill.

Installation of an automatic choke conversion kit is a good economy move. Costing only a few dollars, and relatively simple to install, it provides dash-mounted fingertip control of the carburetor choke valve. With the dash control you use only as much choke as is needed to start the car. Eliminated is the problematic automatic choke with all its cleaning, adjusting, and malfunctioning. It takes but a few cold-engine starts to find the proper hand choke setting. Engine "feel" will let you close the choke much sooner than with an automatic, lean the mixture according to engine needs, and operate at peak cold-engine efficiency. This is very critical if good mileage is to be obtained. A cold engine can use twice the amount of gas as one already warmed, so anything that lessens gasoline consumption while the engine is cold will result in double savings.

You might also consider investing in a hand throttle along with the hand choke. This lets you vary the idle speed from the driver's seat and is helpful during warm-ups where a hand choke is used. In essence, the hand throttle replaces the high-idle cam on automatic chokes and is relatively trouble-free.

Fuel pump pressure regulators are options available at a modest price. Their purpose, as the name implies, is to provide correct, even-pressured fuel flow to the carburetor. They can have some value if your car tends toward overrich fuel mixtures and all conventional attempts to correct it have failed.

Cruise control can be a valuable gas-saving addition to any car. When driving at highway speeds you can set the control to your desired speed and it automatically takes over the accelerator pedal, gently and efficiently applying or easing pressure as conditions dictate. Steady and even speed is maintained with minimum throttle, thus conserving fuel. It is not wise, however, to use cruise control in mountainous country. Add-on electric or mechanical cruise-control units can be purchased for as little as $20.

Many new cars are again offering overdrive units as standard equipment or as an available option. If you are buying a new car and plan to keep it awhile, by all means purchase one with overdrive. It will cut down dramatically on gas consumption and reduce engine wear by allowing the engine to work only a fraction of what it normally would at higher speeds. This is an excellent investment, which will pay for itself in a short time. Overdrive can improve highway fuel economy up to 25 percent.

Instead of overdrive, some automobiles come equipped with a 5-speed transmission, the fifth or highest gear acting much the same way as an overdrive. This is a marvelous built-in gas saver, especially for highway driving. Consider a 5-speed transmission when shopping for a car and enjoy the additional gas-saving benefits it provides.

When having a new muffler installed, make sure it is the proper fit for the year, make, and model car you drive. A muffler that creates excessive back pressure, not allowing free passage of exhaust fumes, is disastrous to gas mileage. It could also cause severe power loss and eventually damage the engine. Low back-pressure mufflers are great for gas mileage.

Install a scavenger tip(s) on the exhaust pipe(s). This is a tip in which the opening points *downward* instead of straight out. The advantage it gives is that the onrushing air has a tendency to suck the exhaust from the pipe, thus easing engine back pressure and allowing for a more rapid discharge of exhaust gases. This has a positive effect on mileage.

Plug-in oil dipstick heating elements and engine block or radiator heaters are valuable aids in combating cold-engine gas consumption. By keeping the engine oil or radiator fluid warm, they reduce cold morning warm-up time, the period of heaviest gasoline use.

If you don't have access to one of the above warmers, place old blankets or rugs over the engine. They will help retain some of the heat. Some cover is always better than none. Preshaped, insulated blankets, made expressly for this purpose, can be purchased through many accessory catalogs or at automotive specialty stores.

In place of, or in combination with, the above heat-retaining methods, a 100-watt household bulb placed under the hood near the battery will help prevent cold from lessening the battery's cranking power. A strong battery means easier cold-engine starts with less wasted gas. The bulb will also guard against fuel line freeze-up and help keep the engine oil a bit warmer.

A dash-mounted timing selector is a low-cost add-on device that will improve vehicle performance and increase gas mileage if used properly. It allows the driver to advance or retard distributor spark according to driving conditions and engine load. Hot and cold engines, idling, acceleration, cruising, city driving, hill driving, and driving with additional weight all (ideally) demand different degrees of timing for maximum economy. Factory distributors are preset, and the ignition timing they produce is, at best, a compromise between highway and city driving needs. With a timing selector you can choose the exact degree of timing needed for each type of driving, and get top mileage for your efforts.

The Ford Motor Company says that a 185–195 degree thermostat will provide better engine economy than one of a lower heat range. Their reasoning is correct, too, because a warmer engine always uses considerably less fuel than a cold one. The lower degree (160–170 thermostats are mostly used when the cooling system contains a pure alcohol-type antifreeze. Since alcohol has a relatively low boiling point, it is necessary to run the engine on the cool side to prevent boil-overs and loss of coolant. It's rare to find the old alcohol-type antifreeze anymore, but if by chance you are still using it, switch to the "permanent" summer-winter type and install a hotter thermostat to obtain better gas mileage.

If you're buying a new car, order one with a lower , more economical rear-axle ratio. It doesn't cost that much more and is a practical investment that will pay for itself in future gas savings. With a lower axle ratio, the rear wheels revolve more times for the same amount of engine work as compared with a standard axle ratio. Thus, you go farther with the same amount of gasoline. In government-sponsored tests of possible fuel-saving methods, a lower rear-axle ratio was recommended as a feasible, economical, and readily available fuel-conserving modification.

Extra-high-resistance ignition wires require more voltage to fire the spark plugs efficiently. If you have a high-intensity coil this is fine, as the coil can deliver enough voltage to overcome the added resistance of the wires. However, high-resistance wiring, in combination with a stock or weak coil, can cause a drop in engine performance and economy. For best mileage, stay with the less resistant, straight-through metal core wires with silicone-rubber insulation; they're more heat-resistant and will last longer, too.

A weak ignition coil won't allow maximum voltage to reach the spark plugs. Low or erratic voltage at time of detonation results in incomplete combustion because the spark supplied by the coil is not "hot" enough to fire the entire charge. The unburned portion goes out the exhaust or is deposited on the cylinder walls; either way, it costs you money. If coil replacement becomes necessary, invest in one of the new high-voltage kinds. They can be purchased for approximately the same money as a factory-type coil and will ensure peak voltage to the plugs under all load conditions. You enjoy better combustion, less engine carbon buildup, cleaner, longer-lasting spark plugs, and increased economy.

A capacitive discharge or an electronic ignition will improve mileage and extend spark plug and valve life. As a rule, one of these improved ignitions requires a fair cash outlay and will pay for itself only if the car is to be driven considerably or kept for a few years. There are many of these on the market, so be selective and choose a brand that has a good reputation.

On cold winter days you've probably seen big trucks (and cars, too) rolling down the highway with the radiator partially covered. By blocking some of the on-rushing cold air that normally passes through the radiator, a warmer and more efficient engine results. Cold morning warm-ups, the most critical period of gas consumption, will also be hastened. A small piece of cardboard or vinyl cloth, placed so that it covers a portion of the front of the radiator, will do the job nicely. Be sure to remove it as weather warms.

If the time ever comes when you have to replace the carburetor, instead of buying an original equipment model consider a smaller one for better economy. It can be easily adapted to your present car and chances are it will cost less than the original. Sometimes the fuel savings gained by switching to a smaller carburetor can be dramatic.

If you are mechanically inclined and won't mind rebuilding the carburetor, mileage-economy kits can be purchased for many models that will do the same job as the exhaust-gas analyzer method described in Chapter 5. Leaner jets, metering rods with longer economy steps, and redesigned accelerator pumps are supplied, along with the basic elements of standard rebuilding kits.

Mileage improvements can run as high as 4 mpg, well worth investment in the kit.

A wire-mesh screen (about 8 grids per inch), inserted between the carburetor flange gasket and the intake manifold, will help "break up" the air/fuel charge and produce a more homogeneous mixture for combustion. Also, on rapid acceleration, excessive fuel and moisture tend to cling to the screen momentarily, instead of being drawn instantly into the intake manifold. This helps stop raw gas from flowing into the cylinders and diluting the engine oil. Make sure that the mesh is sealed tight and no air leaks are present.

A homemade ram air charger, or an inexpensive store-bought unit, will add extra miles to each tank of gas. If you do considerable highway driving, it can be especially beneficial. Dense outside air is force-fed through flexible ductwork directly to the carburetor air cleaner. As the car's speed increases, more air is supplied to the carburetor. Most engines are air-starved at higher speeds and ram air is one way to supply the additional quantities needed for more economical performance.

To make a ram air charger yourself, buy a length of flexible duct hose similar to the type used under the dash for your car's heater blower. Fix one end so the opening will scoop up onrushing air; to either side of the radiator, behind the front grill, is a good spot. Then cut off a section of the air cleaner snout so its diameter approximates that of the duct, and attach the duct to the enlarged opening. If you wish, cut a circular hole in the side of the air cleaner and attach the hose at that point, using small sheet metal screws.

If for any reason the cylinder heads must be removed from the engine (in case of a valve job or some such repair), you might consider having them milled a fraction or install an extra-thin head gasket when they are replaced. Hot-rodders have used this trick for many years as a basic way to increase power and economy by raising the engine's compression ratio. The simplest and least expensive method is to use the ultra-thin head gasket (you have to have one, anyway) instead of one of standard thickness.

Race-car mechanics know of the additional power and performance gains attributed to customized or high-performance camshafts. On the other hand, economy buffs know that cams can be "customized for economy," and they are willing to accept some power loss to gain extra mpg's. However, installing an economy cam is a costly procedure and should be considered only in the event a major engine overhaul becomes necessary.

A dual-intake manifold, popular with hot-rodders and performance-conscious drivers for years, will increase gas mileage by balancing the fuel mixture to each cyclinder. In effect, dual-intakes custom-blend fuel charges for each cylinder, making it possible to use each drop of gas more efficiently.

Designers of high-performance racing cars are well aware that relatively small changes in body shape can result in fuel savings of up to 12 percent at high racing speeds. These improvements in vehicle design will also provide savings at passenger car speeds. Adding a "spoiler" to the underneath of your car is one simple, effective way of altering the vehicle's shape and reducing aerodynamic drag. A spoiler is nothing more than a piece of flat sheet metal or fiber glass that is fixed to the underneath of the car, immediately behind the front bumper. It is fastened in such a way that it covers up the array of protruding engine and front-end parts and negates their effective drag by smoothing out the underneath profile of the car. Mileage gains of 1–3 mph are not uncommon when a spoiler is installed. Indeed, it is well worth the time and few dollars it takes to make one. Glance at the front end of a Citroen automobile to see an excellent example of a built-in front-end spoiler.

If you drive a van or large truck, or do a considerable amount of trailer or camper pulling, consider investing in a wind deflector. These easily attached devices cut wind resistance considerably and can add many extra miles to each tank of gas.

Flexible or "flex" fans reduce the pitch in their blades at high speeds, thereby limiting airflow and lessening engine horsepower requirements. When a vehicle is moving at high speed, it usually gets all the cooling it needs from onrushing air passing through the radiator. By reducing blade pitch at high speed, flex fans allow outside air to do most of the cooling and save gas by easing engine output. A fan clutch can also be used. It "unloads" the fan at high speeds when it is not needed. Both of these are superior to the fixed-type fan. Mechanical slip clutch drive and electrical drive fans are also available, and enable the driver to manually control the fan and use it only when necessary. Figure 8 shows the superiority of flex fans over standard fans at high engine rpm's. Remember, the fan and water pump consume 4 to 5 percent of the total engine-power output.

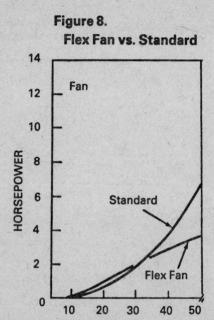

Figure 8.
Flex Fan vs. Standard

Plastic wheels, weighing 2–3 pounds, compared with standard steel wheels at 21 pounds, may soon be available on new cars. They would have the effect of lowering total vehicle weight 100 pounds (this includes the spare, too). Less weight = less gas used, so if you ever find yourself faced with the option, go with the plastic, all things being equal.

Figure 9. Fuel Consumption vs. Elevation

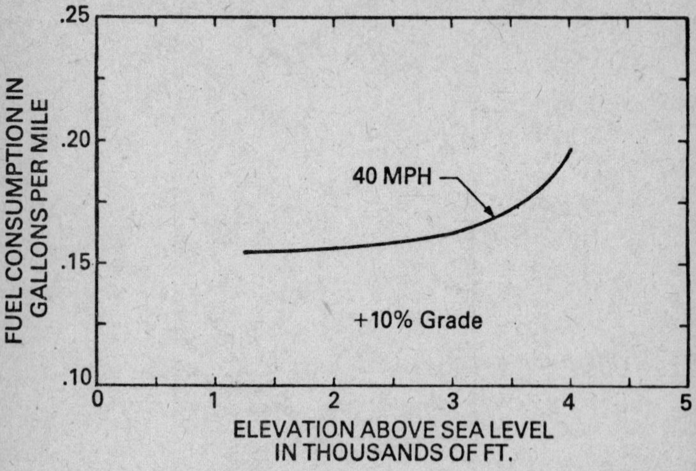

Figure 9 reminds the reader to expect a drop in gas mileage when driving at altitudes above 2,000 feet. The mileage penalty becomes even more severe when elevations above 4,000 feet are encountered.

Figure 10 is a sad commentary on automobile engine efficiency. Of all the energy released as heat when fuel is burned, only 20–30 percent is used to actually propel the vehicle at speeds above 30 mph. A full 70–80 percent is rejected heat and energy consumed by engine friction, accessories, and auxiliaries attached to the engine. No wonder it's a battle to get good mileage!

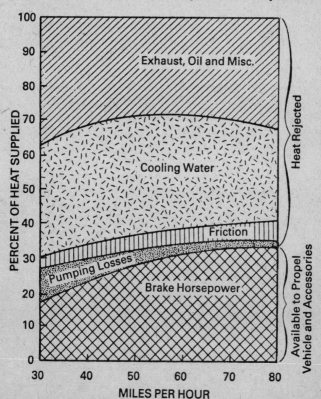

Figure 10. Available and Rejected Horsepower

If you are ever faced with the option, remember that fuel injection is usually more mileage-efficient than carburetion for a given vehicle.

In general, a diesel is much more efficient than a rotary, CVCC, or conventional engine.

A carburetor degasser is a device used to shut off the fuel supply when a high-manifold vacuum is present, such as during deceleration. Up to ten gallons of gas per year can be saved by using this device. A few carburetors, such as the Stromberg NA-Y5G, have one built in. Since the engine doesn't need gas during deceleration, why waste it? That's the theory behind the degasser. It may become standard equipment on all cars in the very near future and if you're lucky, you might live in an area of the country where one can be bought.

A light-colored vehicle will be slightly more economical in hot climates. Why? Light colors reflect the sun; that keeps the car cooler. The air conditioner doesn't have to work as hard to keep the car cool and conversely, you save gas. Remember this when you purchase your next car.

Would you believe a vinyl top can reduce your highway gas mileage figure? It's true. A vinyl top adds extra resistance to smooth airflow over the top of the car. It could cost you ½ mpg. A sun roof acts the same— especially when it is open! Be aware of these popular options that cost you by lowering highway mpg.

Worthwhile Additives, Engine Treatments, and Lubricants

Recent tests conducted independently by the Ethyl Corporation, Automotive Research Associates, and Loughborough Consultants Limited have shown that a 2-12 percent improvement in gas mileage can be expected

if molybdenum di-sulphide (MoS2) is added to the
crankcase oil. "Moly" oil additives have been around
for years, but the recent gas shortage has brought their
remarkable lubricating qualities to the public eye. MoS2
additives are now available at many auto parts stores
under various commercial names. Molybdenum di-
sulphide, when combined with regular motor oil, dramatically enhances its lubricating qualities and cuts
internal engine friction, a major cause of gasoline consumption. "Moly" will also tangentially reduce fuel consumption due to its effect on increased cranking (starting) speed at low temperatures. The reduction in friction is beneficial during the entire "warm-up" period,
when fuel demands are highest.

A U.S. Department of Transportation report shows that
a driver who averages 12,000 miles a year at 12 mpg
with gas cost set at 50¢ a gallon (a low figure) will save
$23 a year on *gasoline* alone by using MoS2 in his oil.
This figure would be much higher with a more realistic
gas price and mileage-traveled figure. Additional savings
are gained through extended oil life, lessened oil consumption, and reduced engine wear and tear. A can of
MoS2 additive with each oil change will pay for itself
many times over. Indeed, it is a "miracle additive" that
you can use right now to start enjoying the extra mile or
two per gallon.

Combining "moly" with the chassis, front-wheel bearing,
and differential lubricants will contribute to the mechanical efficiency of the moving parts by reducing friction
between them. This will, in a small way, also help improve gas mileage.

If you live in a section of the country that gets severe

winters, see if you can find a service station that offers a lighter-grade differential and gear lubricant. Under constant freezing conditions many gears never get a chance to work freely because the lubricant never becomes fluid enough. A lighter grade will warm faster, enhance gear action, and allow the car to move more easily.

If your engine is beginning to use oil and show signs of wear and failing economy, you might try one of the metal-plating, tablet-form treatments that are commercially available. Selling in the $6–$10 range per treatment, these metal tablets are placed in the gas tank, where they eventually dissolve; the suspended collodial particles are then carried to the engine, where they "plate" worn areas. If your engine is not too far gone, they could be of some help in restoring lost power and economy.

Recently, a new engine treatment has appeared on the market that seems to show great promise. It can be recognized easily because the advertising says it contains teflon. Manufacturers claim reduced engine friction and drag, better lubrication, reduced wear and oil consumption, more power, and better mileage. This is a one-time, permanent-type treatment that is added to the oil. Cost is around $15 per can. The treatment is supposed to coat and bond to the engine parts a tough, continuous film of teflon particles so that you have teflon rubbing on teflon rather than metal on metal. The reduction of friction and engine wear is claimed to be dramatic. Although this treatment is still too new to pass judgment on, it seems to hold much promise for better overall engine economy.

Everyone is familiar with the type of oil additive that

comes out of the can looking (and pouring) like pure honey. So many people use this type of oil additive and swear by it that to knock it would be a national crime. So we won't. Can this type of oil additive improve performance and gas mileage? If your engine is worn a bit and using some oil, then these viscosity-extending additives can help increase engine compression with resultant better performance and improved mileage. But according to a recent consumer study, a driver can get the same effect by using heavier (40W–50W) oil in the first place. The choice, of course, is yours.

Weight of the engine oil has been shown to affect gas mileage. A lighter or lower-numbered oil will flow easier and lubricate better when cold than will a heavier, more viscous oil. It also takes less engine power to pump a thin oil than a thick one; thus, a 10W oil will give better gas mileage than a 40W oil. By switching to a lighter oil, especially for winter driving, you allow your engine a better chance to produce top mileage. If you are of the old school (heavy oil protects better) and feel more comfortable with a heavier oil, then multigrade oil is the answer. A 10W–40W will be thin when it is cold and then thicken as the engine warms and demands extra protection. You gain economy by having the thinner, easy-flowing oil for cold-engine temperature when friction and gas consumption reach their peak. Remember, anything we do to improve cold-engine economy pays off in double savings because a cold engine uses twice the amount of gas doing the same work as a warm one. A good-grade thin oil is one way to help cut the cost of cold-weather and cold-engine operation.

What about liquid gasoline additives? Can they help improve gas mileage? There are many of these on the market. and their main value as a mileage aid may be in the fact that they help clean the carburetor and fuel system and lubricate the top cylinder area, resulting in slightly increased engine efficiency.

Synthetic oils—can they save you gas? By reducing internal engine friction a substantial degree over conventional oil. synthetic oil promotes better mileage in the majority of cars it is used in. The mileage improvement varies greatly and is directly related to the condition of each individual engine. Realistically, a driver can expect anywhere from 10–20 extra miles for each 20-gallon fill-up. The high cost per quart (around $4) is the only negative factor. Better performance, less frequent oil changes, better mileage. and prolonged engine life are the plus factors. For our purposes, synthetic oil can and does improve gas mileage.

The 5 Keys to Maximum Mileage - Inspect, Clean, Adjust, Alter, and Repair

The Carburetor

The fast-idle cam adjustment, which controls cold-engine idle speed, should be set on the lowest cam step that allows sufficient cold engine rpm's to prevent stalling. Too high cold-idle speed rapidly eats gas. Keep it as low as possible. In summer, the fast-idle cam screw can usually be set at the lowest cam step for additional gas savings.

Because it may be necessary to increase idle speed to compensate for heavy air conditioner use, remember to turn the idle down when the air conditioner season is over. Without the extra pull of the compressor, engine idle speed can be cut by at least a hundred rpm's.

Some drivers will disconnect and totally bypass the carburetor accelerator pump in their quest for better mileage. Though this will eliminate the gas-wasting squirts of the pump, it leaves the driver with little recourse when fast acceleration is called for. Although you can save some gas by doing it, we don't recommend disconnecting the accelerator pump, because if you have to accelerate in a tight situation you won't be able to do it fast enough.

You can get better mileage by shortening the carburetor accelerator pump stroke. This is done by moving the accelerator pump rod end to a different hole so that it *shortens* the pump stroke. This lessens the amount of gas pumped into the carburetor each time the accelerator is depressed. Expect a slight reduction in top speed with this easy adjustment.

A decrease of 1¼ percent in the air/fuel ratio occurs for every 1,000 foot rise in elevation, so for best economy, carburetor richness should be reset if you move from a lower to a higher altitude, or vice versa. At higher altitudes, where air is thinner and less available, the carburetor must be set *leaner* to compensate for the less dense air. A slightly *richer* setting is desirable at lower elevations, where air is more dense.

If you smell raw gas when starting a cold engine, it usually means that the automatic choke is set too rich. Lean the choke setting to the point of easiest starting. This also is usually the most economical setting. Some newer cars have electrical assists on the choke to help open and close the valve. Check and make certain the choke is functioning properly and the choke valve is fully opened when the engine is warm.

Many carburetors are equipped with a hot-idle compensator. This is a thermostatically controlled valve located in the upper throat of the carburetor and is usually visible when the air cleaner is removed. During long periods of idling with a hot engine (for instance, in summer rush-hour traffic), the fuel in the carburetor bowl can become hot enough to vaporize. These vapors can enter the carburetor bores, mix with the idle air, and be drawn into the engine. This causes an extremely rich mixture and can cause the engine to stall. The hot-idle compensator opens under these conditions, permitting additional air to enter the manifold and mix with the rich fuel vapors, providing a more combustible mixture. Extreme rough idle operation and engine stalling are thus avoided. If your car has a hot-idle compensator and you have any doubts concerning operation of the valve, it should be replaced.

The carburetor float (or floats) should be checked for proper alignment. It should move freely up and down and not scrape or hang up on the sides of the bowl. This will allow excess fuel to enter the bowl. Remove the float and shake it to make sure it isn't filling with gasoline. A gas-filled float will sink, opening the needle valve and allowing continuous wasteful passage of gasoline.

Check the carburetor needle valve and seat for wear and for dirt particles that may be trapped there. A needle valve that doesn't seat properly allows extra gas to "leak" into the bowl and create an overrich mixture, often culminating in severe flooding and wasted gas. Replace the needle valve if it shows any signs of scarring or wear.

On four-barrel carburetors, disconnect the secondary throttle linkage. If you're economy-minded, you won't need the extra power provided by multiple throttling.

A sticking carburetor accelerator pump is an all too common gas thief. If your car hesitates, or the engine almost dies when the accelerator is pushed down, you can bet the accelerator pump is at fault. This becomes more apparent when the gas pedal is suddenly "floored"—the car hesitates, coughs, and then gradually picks up speed. Accelerator pumps are not expensive and are relatively easy to install. They can be purchased singly or are included in carburetor-rebuilding kits.

Automatic choke settings do not have to be as rich for summer as for winter use. Lean the choke when warmer weather arrives, using the leanest possible setting that permits easy cold-engine starting. When the engine has warmed, check the choke valve to see that it is fully opened; if it isn't, the setting is still too rich.

Lowering the carburetor float level approximately 1/16 inch below factory specifications is another way to enjoy a slight increase in gas mileage. Don't lower it more than this amount or you can starve the carburetor.

If your car is a rough idler, don't increase the idle speed to overcome the roughness—that costs you gas. Some of the newer cars, heavily laden with antipollution equipment, are notoriously poor idlers, and unfortunately there isn't much that can be done about it. Set the idle at the lowest rpm where the engine won't constantly stall. If you don't think idle speed has much of an effect on gas mileage, just look at Figure 11. Reducing idle speed from 900 rpm to 400 rpm increases deceleration gas mileage by a whopping 12 miles per gallon and cuts *in half* the amount of gas needed to idle the engine for one hour!

Many recent-model cars have a tendency to "diesel"; that is, the engine continues to run after the ignition is

Figure 11. Idle Speed and Its Effect on Gas Mileage

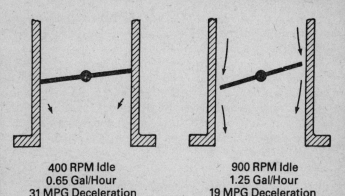

400 RPM Idle
0.65 Gal/Hour
31 MPG Deceleration

900 RPM Idle
1.25 Gal/Hour
19 MPG Deceleration

Courtesy: McDonnell Douglas Corp.

turned off. This not only costs you gas dollars but is harmful to the engine as well. Some cars are equipped with an idle-stop solenoid, usually located near the carburetor, that prevents this condition. The solenoid holds the throttle linkage at idle while the ignition is on; when the ignition is turned off, it closes the throttle to prevent further idling. Proper operation of this solenoid will save you gas dollars. If your car has a dieseling problem, check to see if it has an idle-stop solenoid, and have it replaced if it's not working right.

When replacing carburetor gaskets, make sure that the new ones have the exact same cutouts and holes as the old ones. A gasket that is not exactly correct may plug a vital vacuum or vent passage and adversely affect performance.

Check the idle-air bleed hole on top of the carburetor to be certain it is free of obstructions. A clogged idle vent will hurt mileage by enriching the idle air/fuel mixture.

You can conscientiously follow every suggestion in this book, but if the carburetor air/fuel ratio is too rich, your efforts won't be totally rewarded. The air/fuel ratio must be set "on the lean side" to obtain best mileage. It may be the single most important adjustment you can make to improve gas mileage and it is critical to have it checked and any necessary changes made. The best and most economical way to do this is to take your car to a garage that has an exhaust gas analyzer. This machine will "read" your car's exhaust and tell you if the air/fuel ratio is correct or needs changing. If the ratio is too rich, the garage will probably suggest changing either jets or metering rods to bring it into a more economical range. Some overrich carburetors, when leaned-out, will improve economy up to 5 mpg! Make sure you don't overlean (an air/fuel ratio of 15:1 is usually excellent), because you can damage valves and spark plugs and get erratic engine performance if the carburetor is starved.

When you lean-out the carburetor, make certain the entire ignition system is in good shape because it requires a hotter spark to ignite a leaner mixture (see Figure 12). Exhaust gas analysis, with new jets or metering rods, costs as little as $8 to $10. If the carburetor is overrich, you can save that much in a month by having it leaned. Figure 13 shows how air-full ratios effect efficiency and emissions.

Figure 12. Energy Required for Ignition

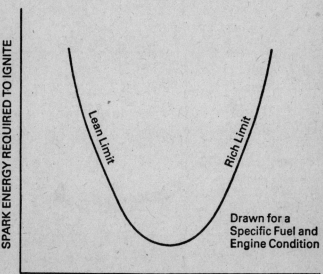

SPARK ENERGY REQUIRED TO IGNITE

Lean Limit

Rich Limit

Drawn for a
Specific Fuel and
Engine Condition

AIR-FUEL RATIO

The Accelerator Pedal

A sticky, balking accelerator pedal will waste gas. The pedal should have a smooth but firm up-and-down movement. A sticking pedal must be pumped and jiggled to be freed, and this costs you gas. Check to see that the pedal is not hanging up on a floor mat or the fire wall hole, and that all linkage moves freely.

If the accelerator pedal is too easy to push, you may be

Figure 13. Effect of Air-Fuel Ratio on Efficiency and Emissions

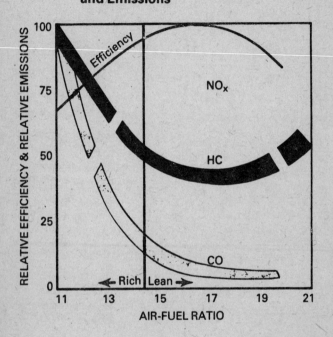

wasting gas then, too. It should offer moderate resistance when pressed. If it pushes too easily, install a stronger or shorter return spring. There are units on the market selling for up to $20 that automatically resist foot pressure, forcing you to use less pedal and save gas.

A shorter or stronger return spring, selling for about $1, will do basicaliy the same thing.

On the other hand, an accelerator pedal that has too much resistance will waste gas also. It is difficult to get good "feel" with too much return-spring tension, and the car will tend to lurch ahead whenever foot pressure is applied. In this case, lengthening the return spring will cure hard pedal and eliminate jerky acceleration.

Some economy-minded drivers place a wooden block under the accelerator to prevent it from being "floored." Although the motivation behind this practice is accurate enough, it can be dangerous. In an emergency, full power may be needed but won't be there because the wood block stops the accelerator. If you now practice this method of saving gas, discontinue it, please. With practice, the egg or apple tricks previously mentioned will give the same results with no danger involved.

Fuel Pump, Filter, and Lines

A properly operating fuel pump is absolutely necessary for top mileage. Excessive fuel-pump pressure will force more gas to the carburetor than it needs and can bend or damage the float arm. Too little fuel-pump pressure, on the other hand, can starve the carburetor, cause missing, overburden the engine, and eventually result in major damage. If you have any doubts about your fuel pump, have it checked for proper pressure (usually around 3 to 5 pounds) at a reliable garage. Heavy raw gas odor may indicate a ruptured fuel pump diaphragm, and is cause for replacement of the unit.

Make certain that your car has a fuel filter. The "in-line"

paper element kind is best. By filtering out foreign particles before they reach the carburetor, it ensures clean, steady gasoline delivery. Change filters every 5,000-10,000 miles or when visibly dirty.

Make sure that your gas cap fits snugly. Any rain or splashed water that might seep into the tank can cause misfire and engine-power loss.

Make sure that there are no leaks in the fuel system. Check the gas tank, fuel lines, fuel pump, fuel filter, and carburetor for signs of escaping gas. Points where metal gas lines are connected to rubber ones are especially susceptible. The juncture of the gas filler neck with the tank is also a likely area. All carburetor screws and nuts should be snug and gaskets in good order. A bad gasket on the carburetor accelerator pump will allow gas to escape through the vent hole and will wet the outside of the carburetor. If you smell gas and suspect a leak, park the car on a clean, dry surface and check for wet spots underneath the car for a clue to the area the leak is in. Remember, too, that sometimes a leak will occur only if the engine is running with normal fuel-pump pressure. Therefore, be sure to check with the engine running also.

Spark Plugs and Wires

Spark plugs, one of the most important items that affect gas mileage, should be cleaned and gapped at least every 5,000 miles to ensure the best possible mileage from your car. If any plugs are found to be bad, they should be replaced immediately and not "nursed" for a few thousand more miles. Just one malfunctioning plug can cut 10 percent off the top of your mileage.

A popular trick used by Mileage Marathon and other economy-minded drivers is to widen spark plug gap an extra .010 over specifications. You will experience some loss of top speed with this maneuver, but it is worth it if you are economy conscious. Some of the newer cars with electronic ignitions have plugs that gap up to .075. On these, don't widen the gap further but keep it at specs.

When cleaning and gapping plugs, make sure that the electrodes are *filed* to a nice sharp, square edge. Spark has a tendency to cling to sharp edges in contrast to rounded ones. Crisp, squared-off electrodes will permit more spark, better combustion, and better mileage. This is one of the main reasons new plugs improve performance.

A rounded feeler gauge is usually more accurate for setting spark plug gaps.

A slightly loose spark plug will rob the engine of full compression and noticeably affect gas mileage. Seat all plugs properly and torque them to specifications. If the plugs require "O" ring seals, they should "give" a little to ensure a proper seal between the plug and engine block. If they are flat, compression leaks may occur.

Spark plug wires shoould be inspected periodically for signs of cracking, burning, wear marks, and oil or grease contamination. Ignition wires are the "arteries" of your engine, and like those in your body, must be in good condition for unimpeded electrical flow.

Keep spark plug wires separated. Grooved plastic separators are usually provided for this purpose. Use them.

Do not tape or tie plug wires together, as this can cause inductive cross-firing across the wires and upset the firing order of the engine.

Plug wires that are too long will build up extra resistance and cut the intensity of the voltage delivered to the plugs, reducing their ability to detonate the fuel charge. If your plug wires are sloppy and longer than necessary, cut them to the shortest practical length that will extend between plugs and distributor cap. The less distance voltage has to travel, the better its quality. Short plug wires improve engine economy. Make certain also that the high tension wire between the coil and distributor is as short as possible.

Sometimes an old or malfunctioning ignition switch can be responsible for hard starting and an overall drop in engine efficiency by impeding electrical flow in the primary ignition circuit.

Spark plugs must be in the correct heat range to get top mileage. If the heat range is too low (cold plug), the plug will foul, cause rough-engine idle, and allow carbon deposits to build up. If the heat range is too high (hot plug), it can burn valves, crack plug insulators, and destroy electrodes. As a general rule, it is best to stay with the type of plug recommended by the car manufacturer. However, additional gas savings can be realized by installing a slightly hotter plug if you do a lot of city driving, a range colder plug if you do mostly highway driving. Let plug condition dictate if you should change, but don't go up or down more than one heat range or plug number at a time. If plugs are heavily carboned and oil-fouled, that's usually a sign that they are too cold and won't allow complete combustion. Metal deposits on the electrodes, cracked insulators, whitish deposits, or eroded electrodes signal that the plug is too hot and a colder range is desirable.

The Distributor

Distributor breaker points that are worn, pitted, dirty, oily, misaligned, or improperly gapped are a major cause of bad gas mileage. Point gap or dwell should be checked at least every 5,000 miles because the distributor shaft rubbing block that causes the points to open and close will wear, narrowing the gap. A new set of points is relatively inexpensive, so if in doubt, replace them.

Some distributors are equipped with an electrical solenoid that allows additional spark advance during starting. If this is not working properly it will make starting hard, if not impossible. Check all connections to and from the solenoid for tightness, and replace the solenoid if there is any doubt about its performance.

On other vehicles, distributor vacuum advance may be controlled by means of a thermal vacuum switch that senses engine water temperature. When the temperature rises, spark is automatically advanced to smooth engine performance. Check all hoses leading to and from the TVS for leaks.

Some cars have a switch located on the transmission that controls spark advance as a function of gearing. When the vehicle is in high gear, the spark advances; when in first or second, it retards. If this switch isn't functioning in the proper gear ranges, it will cut into gas savings. If in doubt, have it checked.

Be certain that the distributor rotor isn't worn, cracked, or dirty, and that it fits snugly on the distributor shaft.

The rotor is the point of spark transference and must be in good condition if a hot spark is to make it to the cylinders.

Check the outside and inside of the distributor cap to make sure it is not cracked, burned, dirty, or wet. Look closely at the electrodes inside the cap for signs of excessive wear. Just a hairline crack, a spot of grease, or a bit of condensation inside the cap can make the spark diffuse over the entire cap surface, causing backfire and engine miss.

A weak or leaking ignition condenser (located under the distributor cap) is an often overlooked thief of good mileage. A malfunctioning condenser won't allow the coil to deliver peak voltage to the spark plugs. Bluing or excessive pitting of the distributor contact points usually indicates a faulty condenser.

The distributor automatic advance must work freely. Its job is to "advance" spark to the cylinders during acceleration and high-speed operation. If a spring or diaphragm in the automatic advance mechanism is broken, engine miss and a noticeable drop in economy will result. The distributor shaft should turn with slight hand pressure and then return to its original position when the pressure is released. The automatic advance must function properly to ensure maximum mileage at lower speeds.

One of the most popular ways to increase gas mileage is to advance the ignition timing 3–5 degrees over factory specs. Factory-set timing must at best be a compromise between the different types of driving encountered throughout the country. Advance it a few degrees and enjoy extra mileage from your car. Although changing the timing can be done rather easily, it is best to let a properly equipped garage handle it. Timing that is advanced too far will do more harm than good, so don't chance it unless the proper equipment is available. "Power tuning," or timing by ear, is out. Through the years it has probably created quite a few problems, and the driver who tries it usually ends up at a garage having the timing reset. Advanced timing, not more than 5 degrees, is a legitimate, effective way of boosting mileage. Economy Run drivers have done it for years. If not overdone, it can boost gas mileage by 1–2 mpg, depending on engine condition, driving habits, and terrain.

For the do-it-yourself tune-up advocate one of the most important items you can be aware of is the *one-way* timing/dwell relationship. When you change the setting of the distributor points (dwell angle), you automatically change the ignition timing; but if you change the timing, the dwell angle *doesn't* change. For this reason it is important that the dwell angle be set *first*; then the timing can be adjusted. This procedure ensures the correct timing/dwell relationship and proper engine operation.

The Coil

Correct coil polarity is essential if you want top economy. This often overlooked area can be a cause of very rough engine performance and mileage loss. A coil with "reversed," or positive, polarity requires up to 50 percent more voltage to fire a spark plug and places a tremendous burden on the ignition system. If you suspect reversed coil polarity, have it checked at a garage or do it yourself by removing an ignition wire from one of the spark plugs while the engine is idling. Hold the end of the plug wire close to the top of the plug while inserting the end of a lead-tipped *wooden* pencil between the end of the wire and the plug. If the spark fires on the plug side of the pencil, polarity is correct; if it fires toward the ignition wire, polarity is reversed. To correct polarity, simply reverse the primary wires at the coil.

The Battery

A fully charged battery is a must if you are trying for maximum gas mileage. No Economy Run driver would be caught without one. A battery in top condition delivers a hotter spark to the entire ignition system, allowing better combustion. A charged battery also signals to the voltage regulator, which in turn keeps generator or alternator output at minimum, thus preventing horsepower loss. Clean terminals, sound cables, and cells filled to the proper level all help maintain a battery at peak efficiency.

Corroded or loose battery terminals can make the alternator work overtime trying to keep the battery charged. Corroded terminals = more resistance to electric flow = poorer electrical current = more alternator work = less mpg.

NOTE: All connections in the primary and secondary ignition circuits must be clean and tight to ensure maximum voltage and proper operation of each individual unit in the circuit.

The Engine

Worn piston rings will cause low engine compression

and excessive oil and gas consumption. New rings require a major overhaul, which these days is very expensive. If you plan on keeping your car for a few more years, the cost of the overhaul will at least be partially taken up in the extra gas savings you will gain from the "new" engine.

Be certain that there are no restrictions in any part of the exhaust system, from the exhaust manifold to the end of the tail pipe. Check for clogged, bent, or dented exhaust pipes, tail pipes, and muffler. Any restriction in the system will cause an increase in back pressure, and total performance will suffer. It is absolutely necessary that the exhaust system be free and clear.

Cylinder head bolts should be checked for tightness to prevent any chance of compression leaks around the head gasket.

Don't forget to check the draft tube located underneath the engine of many older cars. It can become clogged with engine and road dirt, and when restricted, hinders the engine from "breathing" normally and functioning at top capacity.

A bad water pump not only won't provide adequate coolant to vital engine areas but will make the engine work harder to turn it because of bad bearings and scraping rotor blades. To improve water pump life, turn on the air conditioner *only while the engine is idling*. This places less initial strain on the water pump bearings as the air conditioner drive belt is engaged.

Some car owners will remove the air conditioner drive belt during the winter months and go a bit further on each gallon of gas. The only disadvantage here is that the unit should be run occasionally to help keep the seals pliable, and this involves replacing the belt a time or two.

In very cold winter climates you could try running without a fan and enjoy more gas savings. However, keep a prudent eye on the temperature gauge, and if the car is running too hot, replace the fan belt. If you try this, keep the fan belt in the trunk in case it is needed. By removing the fan belt you not only bypass the fan but also the water pump. Combined, these two units use up to 8 horsepower when engaged, so it is obvious that considerable gas can be saved by disconnecting them if weather allows. Figure 14 shows water pump/fan power requirements.

All engine drive belts should be adjusted to proper tension. Air conditioner, power steering, power brake, air pump, supercharger, and fan belts should be checked periodically. Belts that are too tight will harm bearings and exert a negative influence on gas mileage because the engine will have to work harder to overcome the extra belt and bearing friction.

Burned or sticky valves lower engine compression and can cause extreme power loss. A sticky valve can sometimes be freed by adding a can of top cylinder oil to the gas or crankcase oil. If this doesn't work, the valve is probably burnt and must be replaced.

Mechanical valve lifters should be adjusted every 10,000 miles. This is especially important with some of the small 4-cylinder engines. A gradual loss of power and economy will occur if valves are not adjusted at regular intervals.

Figure 14. Power Requirements—Water Pump and Fan

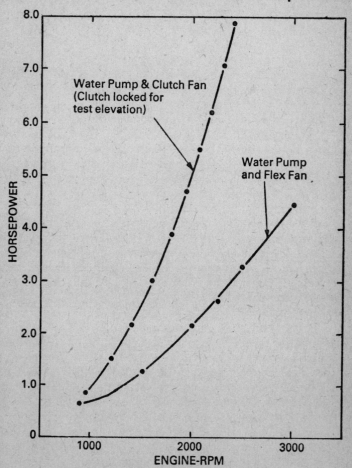

Although hydraulic valve lifters do not require adjustment, you can extend their life by taking it easy the first minute or so after starting the car. This gives the oil a chance to reach the lifters while the engine is still under light load.

Intake manifold leaks caused by any of the following can upset the air/fuel ratio and result in rough idling and poor economy: loose manifold connections or leaks occurring in intake manifold vacuum lines or at the carburetor flange; loose manifold nuts; distortion or misalignment of gasket surfaces at the intake manifold and carburetor-attaching flange; damaged or improperly installed gaskets; a leak at the juncture of the carburetor and the throttle rod where the rod may have worked loose the vacuum seal. To domonstrate the importance of a sealed intake manifold (no leaks), McDonnell Douglas Corp. was able to improve the fuel mileage of some new automobiles with these problems by as much as 4 miles a gallon simply by resurfacing and properly mating the carburetor to the intake manifold! as you can see, air leaks into the intake manifold can be costly.

A large leak, such as a vacuum hose that has become disconnected, may make itself known by a loud hissing sound or can be detected by visual inspection. Less obvious leaks can be located by squirting a little gasoline or household oil in and around suspected areas while the engine is idling. If there is a leak, the gas will be drawn through it and into the engine, where it will cause an increase in engine speed as it is burned along with the regular fuel. Don't forget to also check hoses to vacuum-operated units such as the windshield-wiper motor, windshield washer, and distributor advance.

The Transmission

Automatic transmissions should be adjusted to shift at factory-recommended intervals, or a mile or two lower, for most economical driving. Too much time in low range foolishly wastes gas, and too short a time will cause the car to labor once it has shifted to high. Check the owner's manual or ask a qualified service person at what point your car should shift. If it is shifting too early or too late, have it adjusted. The cost is minimal and gas savings will soon pay for it.

A "slipping" automatic transmission will eat gas rapidly because much of the power sent to the rear wheels is lost due to slippage. The transmission should shift smoothly and surely. If it hesitates, jerks or whines, or slips its way into a higher gear, have it checked. Gas which should be propelling the car is instead being wasted.

By the same token, a slipping clutch in a standard-transmission vehicle will also use extra gas. A slipping clutch is never fully engaged, and power is lost throughout the entire drive train. If the clutch plate isn't worn too badly, a minor adjustment will cure this. Three-quarters' inch of free play in the clutch pedal is usually the best adjustment.

The Air Cleaner

A clean air filter is another "must" for peak gas mileage. When the air filter becomes clogged and dirty the volume of air that is capable of passing through it is reduced, air-starving the carburetor and forcing it to use more gas. Replace the air cleaner *when it is dirty,* disregarding the length of time it has been in use, be it a week or four months. Remember, in areas where there is a high percentage of dirt roads and blowing dust is common, the filter element will tend to clog up fast and should be replaced more frequently.

An easy way to get more efficiency from the air filter element is to simply rotate it 180 degrees. This brings the cleanest part of the filter around to the point where air is drawn through the horn. By doing this, the "unused" part of the filter is brought to the point of air contact and is able to filter more effectively and with less restriction, thus increasing the volume of clean air available for combustion.

Drilling a line of holes around the circumference of the air cleaner is another way of supplying extra air to the carburetor. Some drivers will even go so far as to cut the entire side out of the cleaner, leaving only the top and bottom and exposing the entire surface of the air filter element to outside air.

Instead of the above method you can "dehorn" the air cleaner. This consists of nothing more than cutting off a section of the air cleaner horn or snout so that it has a wider opening, permitting more air to enter.

Look inside the air-cleaner snout to see if it has a diverter valve. This valve monitors the amount of pre-heated manifold air that can be drawn into the carburetor via the air-cleaner diverter tube. It must be free-moving to ensure correct amounts of outside (cold) and manifold (warm) air. If stuck in the closed position, the engine will be robbed of extra air it should be receiving.

The Wheels

Wheel bearings that are too tight will create excessive friction and drag at the bearing/sleeve juncture and won't allow the wheels to turn freely. The adjusting nut must be torqued to specifications.

Improper front-end alignment can cost you an extra 1 to 2 gallons of gas per tankful. Bad alignment can cause the car to "pull" heavily to one side or the other, or it may force the front wheels to turn in and literally "plow" the road. A car with bad alignment will want to go one way while the wheels try to go another. This constant battle between forward movement and drag created by the misaligned wheels costs you more than just gasoline, because tire (and steering component) wear is dramatically hastened.

Dragging brakes are too often a factor in poor gas mileage. When the brake shoes constantly rub against the drum, unnecessary friction is built up at the wheels. The engine must then develop additional horsepower to overcome the friction-created drag. Suspended wheels should rotate easily and be free of any scraping sounds when spun by hand. (Note: With disc brakes there will probably be some sound, as friction pads are always lightly touching the revolving wheel.)

An emergency or parking brake that is adjusted too tightly will cause the same conditions as dragging foot brakes. Wheels should be completely free once the brake is released.

Many drivers forget that rear wheels can get out of alignment too. This is especially true of many foreign makes with the engine in the rear. If you notice the rear tires starting to wear on the inside, that's usually a sign that they are out of line. In most cases, rear-wheel alignment can be done only by a new car dealer's service department.

Emissions-Control Equipment

On post-1966 automobiles the entire emission-control system usually consisted of a simple PCV (Positive Crankcase Ventilating) valve. This simple check valve meters recycled crankcase fumes and allows them to be drawn into the intake manifold, where they are "reburned." If this valve becomes clogged or stuck, the air/fuel balance is upset and rough idling and poor performance result. Replace the PCV valve if in doubt.

On newer cars the EGR, or Exhaust Gas Recirculation valve, should be checked to make certain it isn't stuck, because it controls the amount of burned gases from the exhaust system that are recirculated back into the engine. A frozen valve can result in a mileage drop and poor idling.

Newer cars equipped with the SDS (Spark-Delay-System) pollution control have a Spark Delay Valve in the exhaust-emission-control system. It is important that this valve be checked and replaced if it is not functioning. Otherwise it may become a major cause of poor mileage if it is not working properly, because it directly affects the distributor-advance mechanism.

Check your air cleaner to see if it has a small filter element where the crankcase vent hose attaches (if your car is so equipped). This filter must be cleaned and replaced periodically for proper operation of the emission-control system.

Don't tamper with any emission-control system in the hopes of getting better mileage. Besides being illegal in most states, it can do more harm than good. In a recent EPA test, private service garages were asked to try to get better mileage from a number of vehicles equipped with emission-control devices. How did they do? EPA noted that "emission control system tampering is more likely to hurt fuel economy than to improve it. Such tampering virtually always makes emissions worse, and can cause deterioration in engine durability. Regular maintenance according to manufacturer specifications improves both emissions and fuel economy." Figure 15 shows the results of this test.

Figure 15. Effects of Private Garage Tampering

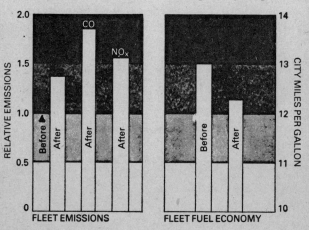

The Body

If two cars are exactly alike in every respect, but one is carrying four passengers while the other has only the driver, the one with the extra passengers will suffer a 6–12 percent decrease in gas mileage due to the weight of the extra occupants. Weight has a direct bearing on gas mileage. For every 100 pounds of added weight, mileage decreases from 1–6 percent, depending on the size of the car. Check the trunk and rear seat and remove any unwanted articles—it costs you to lug them around.

Figure 16. Weight versus Fuel Mileage

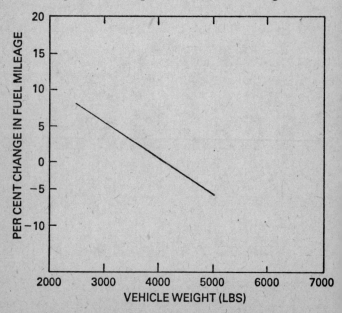

Figure 17. Weight vs. Engine Size

1800-Pound Vehicle 85-Cubic-Inch Engine	5000-Pound Vehicle 429-Cubic-Inch Engine

With 100 Pounds Added Fuel Mileage Decreases 2.1 MPG	With 100 Pounds Added Fuel Mileage Decreases 0.28 MPG

Courtesy: McDonnell Douglas Corp.

Figure 17 dramatizes the effect of weight versus fuel mileage. It shows that additional weight has a more pronounced effect (in terms of mileage) on a small car with a small engine than it does on a large car with a large engine. But with either car it's still you, the driver, who pays the penalty.

It has been demonstrated that a 10 percent decrease in aerodynamic drag (the car's susceptibility to wind resistance) of an intermediate-size car results in a 5 percent increase in gas mileage at 60 mph. A cleaner, more aerodynamic profile means better gas mileage, and although you can't do much to alter the overall shape of your car (spoilers are an exception), you can clean up what you have. Luggage and ski racks should be removed when not in use. Pennants on antennas, bug deflectors, too many or oversize mirrors, big hood ornaments—yes, even mud flaps—create air turbulence and "hold back" the car. Clean up the outside of your car for improved mileage.

If you live in a state that requires a front license plate, position it so that it doesn't contribute to the car's aerodynamic drag. If it hangs down into the air stream, move it up and fasten it to the bumper.

Keeping your car washed and waxed is another effective way to lessen aerodynamic drag and gasoline consumption. A smoother surface will "shed" wind better, enabling the car to move with less power output.

What's in store for drivers in the near future? A recent study conducted for the Department of Transportation and the EPA had as its goal a 43 percent improvement in fuel economy for all passenger vehicles by the year 1980. The study concludes:

The most cost-effective approach—with medium technical risk . . . involves use of a light-weight diesel engine, combined with a four-speed automatic transmission, a torque converter equipped with a lock-up mounted in a light-weight improved body, and a chassis equipped with radial tires.

6 How Regular Autombile Maintenance Yields More Miles Per Gallon

Every moving part of your car has an effect on gas mileage! A car that is well lubricated delivers better mileage than one that isn't. Follow a pattern of regular-interval lubrication as recommended by the manu-facturer—more often under extreme conditions. Ball joints, universal joints, differential, and other lube points work better when properly lubricated. Remember that power loss due to friction occurs at every bearing and gear in your car. Regular lubrication will keep the loss at a minimum and save you gas.

Front-wheel bearings should be cleaned and repacked with proper-grade lubricant at least every 10,000 miles. Wheel bearings are a critical point at which the moving wheels meet the stationary car frame—an area of ex-treme friction buildup. They must be clean, well-lubricated, and free of "burns" or pits if friction is to be minimized. A bad bearing can usually be detected by a loud intermittent rumbling or grating coming from one of the front wheels. It should be replaced immediately.

Clean the oil breather cap (the one you remove to add oil) at least once every 5,000 miles, and more often under severe dust conditions. A clogged cap affects the engine's air-drawing ability, and if it is not clean, will cause it to use more gas. Some cars have two, so be sure to clean both. Parts cleaner, kerosene, or gasoline will do the job.

A clean engine on the outside as well as the inside goes hand in hand with better mileage because a clean engine runs better and is less likely to develop problems.

Clean oil = clean internal engine = better performance. It is essential that engine oil and filter be clean. Change the oil and filter at prescribed intervals (2,000–5,000 miles), and more often if weather conditions dictate. Oil must be clean to be able to reach and lubricate vital engine parts. Dirty oil carries in it the tidings of early engine failure. Oil is the lifeblood of the engine, so don't skimp when it comes time for a change. Invest in a high-quality, reputable brand; it's worth the few extra nickles per quart, and is excellent insurance against future engine problems.

Maintain engine oil at correct level. The engine needs all the friction-fighting protection it can get. Low oil level won't provide adequate lubrication, and engine efficiency will drop. Keep the oil *full*.

The same applies to the transmission fluid. Check it often and keep it full for best performance.

The heat riser is a thermostatic valve located at some point along the exhaust manifold. It must move freely to ensure maximum performance from your car. During engine warm-ups the valve remains closed, forcing warm exhaust gases to stay a bit longer in the manifold and thus hasten the warm-up; then, as the engine warms, the heat riser opens and permits normal passage of exhaust gases. A sticking or frozen heat riser can cause very rough idling and poor response from a cold engine. Spraying the valve with penetrating oil while giving it a few taps with a hammer will usually free it. No tune-up should be considered complete unless the heat riser has been checked to be sure it is free.

The importance of regular-interval tune-ups cannot be overemphasized. If you take your car to a mechanic, be sure it is tuned at least twice a year, or a minimum of every 6,000 miles. Look for a shop that offers a *dynometer* tune-up. This instrument can simulate actual road driving conditions and tune the car accordingly. The result is a more accurate and lasting tune-up.

If you do your own tune-ups, it's a good idea to work on the distributor, plugs, and timing first. After these are satisfactory, then do your carburetor adjustments. Why? Because ignition can affect the carburetion. For a better, truer tune-up, adjust the carburetor last.

7 Tires - Just Between You and the Road They Can Save You Gas

Stay away from the popular "wide-track" tires if you want top mileage from your car. They are strictly performance and "looks" tires and will cut mileage. The narrower the tread *width* of a tire, the better its gas-conserving qualities. Narrow-tread tires produce less friction at the road surface and thus create less rolling resistance, making it easier for the car to move. If it were possible to equip cars with bicycle tires, gas mileage would be phenomenal. Remember: The narrower the tread, the better the gas mileage.

Adding an extra 3–5 pounds of air to each tire won't noticably affect the riding qualities of your car but it *will* increase tire life and gas mileage by reducing the rolling resistance and friction generated by overly soft tires. Air is free, so don't be afraid to use a little more of it and enjoy an extra 1–2 miles per gallon.

Be sure to add more air to the tires when the weather turns colder. A 10 degree drop in temperature will reduce tire pressure by 1 pound. The difference between summer and winter tire pressures could be as much as 8 pounds less. This could cost you 2 mpg if not corrected.

Remember to check tire pressure often. A slow leak of just a few pounds a week can silently rob you of gasoline and hasten tire wear.

Recent fuel-economy tests conducted by Firestone showed overwhelmingly that radial tires improve gasoline mileage when compared with bias ply or nonradial tires. Improvements in gas mileage ranged from 7–10 percent depending on the speed of the test cars. Translated into miles per gallon, this means that a car now getting 15 mpg with conventional tires would improve to 16–16.5 mpg simply by changing to radial tires. Projecting this further, you would use 205 *less* gallons of gasoline if you drove the radials a full 40,000 miles, the usual guaranteed mileage. At 65¢ per gallon this would mean a savings of $134 in gasoline alone—surely an impressive figure. Be sure to consider radial ply tires when the time comes for replacement, and enjoy the added savings and safety they provide. If you do a significant amount of highway driving, radials can be especially important to you.

Figure 18 demonstrates the superiority of radial tires over bias and bias-belted types. Rolling resistance, a major factor in gas mileage, is significantly less with radials.

Using oversize-diameter tires on the rear of your car has the same effect as a high-speed rear-axle ratio. By using larger tires on the rear, more distance is traveled with each revolution of the rear axle, and slightly better gas mileage is realized. You may have to contend with a slight power loss in lower gears.

Figure 18. Effect of Tire Construction on Rolling Resistance

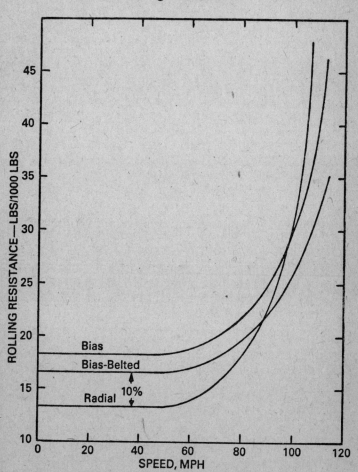

Although it would be a little impractical, if you really wanted top mileage and would go to extremes to get it, driving on treadless tires would give you a big head start. Look at Figure 19 to see the effect tread removal has on rolling resistance. It cuts it in half! So if you're ever in a mileage marathon, "baldies" will up your gas mileage considerably.

If you live in a rainy climate, it will pay you to shop for tires with extra-wide grooves in the tread. These "rain" tires lessen rolling resistance caused by water buildup on the road by channeling away the water more efficiently. They are definitely a mileage "plus" in rainy areas.

Tires that are not balanced properly tend to hop and shimmy, creating additional rolling resistance and drag that the car must overcome. Well-balanced tires are a prerequisite if top gas mileage is desired. Be sure to balance the rear tires as well as the front, because they tend to exhibit similar characteristics when unbalanced. Then, when rotating tires, you won't have to bother balancing the rear ones.If the spare is used in the rotation, it should be balanced also. When balancing new tires, be sure to use the valuable tip given below.

Whenever you purchase a new set of tires, *don't* have them balanced right away. New tires should be driven a few hundred miles so that they can acquire a "set," each tire conforming to its own particular traits. If no balance weights are present to affect the tire during the break-in period, the set will be truer. Drive a few hundred miles on the unbalanced tires, and *then* have them balanced—it will be more accurate and longer lasting. You will gain many extra miles of tire wear, and economy will improve because the tires generate considerably less rolling resistance.

Figure 19. Effect of Tread Removal

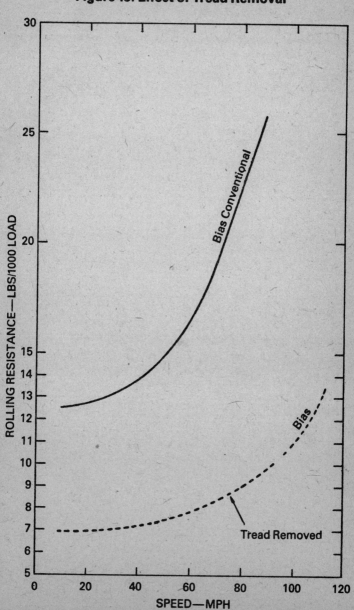

Snow tires should be installed no sooner than necessary and removed as soon as weather permits. The deep cleats on snow tires are designed for traction in snow or mud and generate more friction and rolling resistance than conventional tires. You pay a 1–3 mpg penalty when driving with snow tires, so as soon as they aren't needed, take them off.

Get those snow chains off the car—use them only under the severest weather conditions. Driving with chains on the rear tires is like driving with an extra 1,500 pounds of weight in the car.

Sometimes it is practically impossible to get a tire to balance because of an imperfection in its basic structure. This is occasionally the case with so-called "blems." Too much rubber at points along the tread surface make the tire uneven and a bear to balance. If you are stuck with this type of tire, try having it "trued." Tire trueing is available in most larger cities and will probably cure the existing problem. Thin layers of rubber are peeled off the tire until the surface is even and the radius equal in all directions. It could save you buying another set of tires and will improve mileage to boot.

Most likely your car is your primary means of transportation to and from work. Why not set aside one day a week as public transportation day? Ride the bus, trolley, or train or use other modes of public transport to get to and from work. And have you ever considered owning a bicycle?

Accessories Negatively Affect Gas Mileage

Any time you use a battery-powered accessory you use extra gas. Radio, tape deck, cigarette lighter, power seats, power windows, interior lights, heater and blower fan—all use electrical current. When one of the accessories is used, the alternator/generator is activated to restore to the battery what is being drawn off. This extra work of the generating unit is paid for with gas burned to provide the additional horsepower needed to turn the unit. Use accessories sparingly. Remember that accessories cost you gas in two ways: 1) it takes extra horsepower to run them; 2) their bulk adds considerable weight to the car and you must pay with your gasoline dollars to haul that weight around.

Use of battery-operated accessories while the engine is *off* still must be paid for with gas burned once the engine is started again. The alternator must go to work again, charging the battery. When the ignition is off, be certain lights, radio, and all other battery-operated accessories are also off.

Roll up power windows *before* stopping the car. This prevents running the engine while waiting for the windows to roll up—a small item to be sure, but it saves a few seconds' idling time.

Remember to turn off the air conditioner whenever you park the car. This will save a few drops of gas when the car is restarted because the engine won't have the extra drag of the compressor lowering cranking efficiency.

If you don't think you pay a gas-mileage penalty for running your air conditioner, just glance at the comparative gas mileages of the car with A-C on, and A-C off in Figure 20. As you can see, the penalty may run as high as 4 mpg at speeds of 20–40 mph. Be selective when using the air conditioner. (Use the "vent" position when full cooling is not required.) Don't run it when open vents or cracked windows will do the same job—save the air conditioner for the really hot days and you can gain up to 4 extra mpg while it is off.

It's obvious that you can save gas by joining a car pool. You'll cut your gas bill by the number of people in the pool, so if you are not already pooling, try to find one you can join. Many cities offer a free service of computer matching for persons desiring to join a car pool. The place you live and work, and the times you leave are matched with others in your area, and compatible pools are arranged. Check to see if your city offers a car-pool assistance program. The American Petroleum Institute says that the U.S. could save upwards of 33 million gallons of gasoline *per day* if we could add just one person to the average commuter passenger load.

A recent study by the Federal Highway Administration

Figure 20. Road-Load Fuel Economy

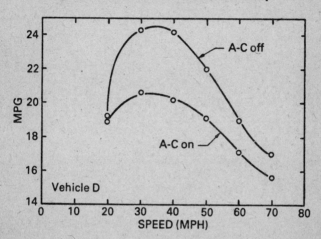

shows that if you drive 10 miles each way to work in an average-size car, the yearly cost for driving is nearly $1,000. If instead of driving alone you decide to join a five-man car pool, you can save about $650 of that total. This money-saved figure includes gasoline, parking, insurance, and repairs. That's quite a saving! Find out if your company has a van pool. This is an efficient, entertaining, and economical way to get to your job.

9 Other Conditions that Cost You Miles Per Gallon

The type of road surface you drive on has a direct relation to the mileage you get. A loose gravel surface can cut mileage by a full third, a muddy road even more. The smoother and firmer the road surface, the better your mileage will be. Stay with good roads when you can.

Some conditions that can influence fuel economy are listed below, with their economy penalties based on steady speed cruising at about 50 mph.

Remember that the berm of the road—that is, the angle at which each side of the road slopes away from the center—may differ from state to state. If upon changing locales you notice that your car now pulls to one side, have it checked for alignment. It may have to be reset to conform to the new roads you are driving.

Avoid driving on wet roads unless it is absolutely necessary. The car must put forth extra effort to "push" itself through the water on the road, and you can lose 1 mpg

Figure 21. Effects of Road Conditions and Environment

Road Conditions:	MPG loss
Broken & patched asphalt	15%
Gravel	35%
Dry sand	45%
3% Grade	32%
7% Grade	55%
Environment:	
18 MPH tailwind ....(19% gain)	
18 MPH crosswind	2%
18 MPH headwind	17%
50°F ambient temperature	5%
20°F ambient temperature	11%
Altitude (4000 ft)	15%

Courtesy: Environmental Protection Agency

on wet surfaces. Government researchers on fuel economy recognize this fact and terminate any economy tests "if the pavement becomes damp enough that the car leaves visible tracks."

If rain on the road can cost you 1 mpg, snow is even worse. As little as one inch of snow on the road is a formidable barrier for the car. It's like constantly climbing a hill, and that requires a lot of extra gas. If you must drive on snow-covered roads, be prepared to have your mileage drop considerably.

Scrape all snow and ice off the car before starting out. Wet snow can really weigh a car down. Don't use extra gas lugging it around.

The colder the ambient or outside temperature, the worse your mileage will be. This is due to the fact that cold air is more dense and increases aerodynamic drag. Reduce your top speed a bit during winter months and you will offset any mileage loss caused by the colder air. Figure 22 gives fuel consumption versus air-temperature curves for a car traveling at 50 mph.

Figure 22. Fuel Consumption vs. Air Temperature

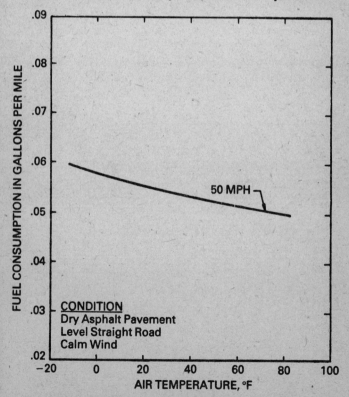

Strong wind will have either a positive or negative effect on gasoline consumption, depending on the direction from which it comes. To illustrate: U.S. Department of Transportation studies have shown that a full-size test car got 13.9 mpg at 70 mph with no wind. The same car, traveling at the same speed but with an 18 mph *headwind*, got only 11.6 mpg. A car going 70 mph with an 18 mph *tailwind*, got a gas mileage of 16.6 mpg, a whopping 5 mpg increase over the car that had to buck the headwind.

Use the wind to your advantage. Strong headwinds should be a signal to slow down so you can reduce the effect of the extra wind resistance on the car. Don't fight headwinds.

A tailwind is a sign to increase your speed slightly and take advantage of the additional "push" the wind is providing. You'll be able to go a bit faster and won't have to pay a gas penalty for doing so. Let the wind do some of the work and e-a-s-e up on the accelerator. Keep an eye out for changing wind directions and monitor your driving accordingly. Swaying trees or bushes, paper blowing across the road, smoke in the air—all give clues to wind direction and velocity. Use the wind when you can—it's free.

INDEX